The Rivers and Streams of Dublin
(City of Dublin, Fingal, South Dublin and Dún Laoghaire-Rathdown)

with notes on the rivers and streams of the cities of Cork and Belfast

2024

Joseph W. Doyle

The Rivers and Streams of Dublin (City of Dublin, Fingal, South Dublin, Dún Laoghaire-Rathdown)
Published by Rath Eanna Research, Dublin, Ireland.
Copyright © Joseph W. Doyle, 2024.

Photographs copyright © Joseph W. Doyle 2006-2024. All rights reserved. The moral rights of the author have been asserted. (Certain historical material - including expired Crown and Irish Government copyright items - is no longer subject to copyright; a few pictures have been opened to online "commons" projects.)

Third edition, 1 February 2024.
ISBN: **(HB) 978-1-9999497-6-1** (HB)

First and second editions 2022

Over 320 illustrations (incl. a few historical scenes, and a few maps plus the back cover county map).
CIP Data: 210 x 148 x 12 mm, iv + 101pp + illustrated covers.

Succesor to **The Rivers and Streams of the Dublin Region**
   First edition, ISBN: 978-0-9566363-8-6   3 printings, 2018 (and one private supplementary release)

and **Ten Dozen Waters: The Rivers and Streams of County Dublin**
   Eighth edition, ISBN: 978-0-9566363-7-9   September 2013
   Seventh edition, ISBN: 978-0-9566363-6-2   3 printings, February - May 2013
   Sixth edition, ISBN: 978-0-9566363-5-5   3 printings, October 2012 - January 2013
   Fifth edition, ISBN: 978-0-9566363-4-8   3 printings, February - May 2012
   Fourth edition, ISBN: 978-0-9566363-3-1   3 printings, October - November 2011
   Third edition, ISBN: 978-0-9566363-2-4   September 2011
   Second edition, ISBN: 978-0-9566363-1-7   3 printings, 2011
   First edition, ISBN: 978-0-9566363-0-0   4 printings, 2008-2010

Images
Front cover (clockwise from top left): River Tolka where Fairview, Ballybough and North Strand meet, from Annesley Bridge / footbridge (modern work replacing a historical bridge) over the River Dodder at Firhouse, just downstream of the City Weir at Balrothery - two small streams from the mountains join the river on the right hand side / Crinken or Wood Brook Stream, Shankill / St Bride's Stream (aka Cabinteely or Foxrock Stream) passing through a basin on its way to Cabinteely Park / the "Lipstick Stones" crossing of the River Dodder near Butterfield Avenue, Rathfarnham, alongside Bushy Park / pond on a tributary of the Whitewater Brook on Howth, just before that streamlet meets the main Brook / the Liffey in the heart of old Dublin, looking towards Fr Mathew Bridge (almost on the site of the original, and for centuries only, bridge in Dublin – the Bridge of the Dark(-Haired) Foreigners, or simply The Bridge of Dublin) / bridge over the Shanganagh River near its mouth.

Page iii (Contents) (from left to right): The Dodder flows into the Liffey, taking in the Grand Canal at the Grand Canal Docks / the Dundrum Slang / Slang River (a Dodder tributary) "steps down" by the M50, heading towards Dundrum village.

Page iv (from left to right): A stream from Rathingle south of Swords winds its way along near Main Street, to meet the Ward River near St Columb's (Colmcille's) Well / the Naniken River passes an ornamental bridge in St Anne's Park / the Camac River approaching Clondalkin.

Contents

Introduction	1
Rivers and Streams of Northern Dublin	2
The River Tolka and its system	30
The River Liffey and its system	38
(incl. R. Camac, p. 52, R. Poddle, p. 58)	
The River Dodder and its system	65
Rivers and Streams of Southern Dublin	78
The author	*89*
End notes	90
(incl. notes on waters of Cork city, p. 91, Belfast, p. 92)	
Partial bibliography	93
Biographical notes (C.L. Sweeney, C. Moriarty)	*95*
Index of watercourse names (natural, artificial)	96
List of Liffey tributaries and reaches (Wicklow, Kildare)	97
Supplementary photographs	*98*

Illustrations

Contemporary photos	Cover, throughout main text, and pp. 98-99	
Historic images, incl. maps		*100-101*
Map of County Dublin (1830s, for Lewis)		*Back cover*

Dedicated to all caring for Dublin's waters, in the local authorities (incl. LAWPRO) and the Office of Public Works, Irish Water / Uisce Eireann, the EPA and volunteer groups / NGOs.
In memory of Clair L. Sweeney and Christopher Moriarty.

Notes to the text
The listing runs north to south, circling Howth Head clockwise on the way.

Naming convention: The name of each river or stream with a Dublin presence, at its primary appearance in the text, is in **bold** typeface, and alternative names at first appearance are in ***bold italics***. here a watercourse is named *only* in bold italics, its status is in some way not fully clear. Rivers and streams with no presence in Dublin, such as many of the upper Liffey tributaries, are <u>underlined</u>. Where known, names in Irish are included in bracketed italics; most must have had names *as Gaeilge* but most of these are long lost.

Map: The back cover map is from the 1830's, issued with Lewis's *Topographical Dictionary of Ireland* (1837). It traces more than three quarters of the watercourses named here, with a couple having course differences; it also shows a range of the "now nameless." On the left-hand side it depicts three exclaves (around Tipperkevin and Ballymore Eustace, and near Dunlavin and Timolin) which were then within County Dublin, although wholly detached from the main body of the county, lying between Wicklow and Kildare. These "islands" - and several like exclaves in other parts of Ireland - were "tidied up" during the 18th and 19th centuries, and the three shown here were removed from Dublin, under an Order in Council, within a few years of the map's publication. The main additional watercourse found in this vicinity is the River Greese or Griese, a tributary of the River Barrow, and in turn its tributary, the Bothoge; the Greese flows on through the "Quaker village" of Ballitore with its Shackleton-founded school. The map has scales in both statute ("English") miles, and the longer, but no longer used, Irish miles.

Thanks... to other researchers in this area and all who have kept memories of Dublin waters alive. And to the Water / Drainage & Archives staff in past and present local authorities. And research sources - due to travel, I had limited chances to use most facilities, but public library systems (with special mention to the city's second-busiest library, at Raheny), and especially Local Studies units, as well as the National Library of Ireland (particularly the Manuscript Room) and the libraries of Trinity College Dublin, the Irish Architectural Archive, and the Royal Society of Antiquaries of Ireland, have helped. As a member of the RSAI for years, I've found their online research materials and access to JSTOR of great value in preparing later editions. Beyond the always excellent *Ordnance Survey* materials (and the clever "layered viewing tool", 2012-2020, sadly missed), the *East West Mapping* (a fine Clonegal company) maps of the Dublin / Wicklow mountains and southern Dublin are great sources - I have yet to see comparable map-related modern work on placenames. Special thanks to my patient wife Luiza, and growing daughters, and to my parents and teachers, and all other supporters, including Ciaran and all the Gilchrists, Pat Crimmins, the Barrys and the Dohertys. Last but not least, thanks to all who have bought, read, gifted or recommended the book, and especially those who got in touch. I especially liked the "find the river" trails sent from Templeogue and reports of quiz uses - and am also delighted to see the book used as a reference on Wikipedia - and I hope that articles can be added there for more of the interesting watercourses).

Plans: The book passed the 15th anniversary of the project between editions, with good interest and sales, and some hopeful work around rivers around the city, from cleaning (especially great work in the Dodder system) to the still-promising chance to daylight a bit of the Naniken. I still hope for time to work on a Frequently Asked Questions page, maybe some quiz ideas, and eventually improved mapping.

The Rivers and Streams of Dublin (City of Dublin, Fingal, South Dublin, Dún Laoghaire-Rathdown)

Introduction

This book presents the rivers, streams and streamlets of the full traditional County of Dublin (briefly the *Dublin Region*, and now holding four modern county-level administrations – the city plus Fingal, South Dublin and Dún Laoghaire-Rathdown) from deep city to rural corner. There are more than 135 clearly-named natural watercourses in the county (and over 160 if we consider all with possible names) but few can name more than a dozen offhand. Yet you may find busy waters just over a wall, crossing a beach, passing a familiar shopping centre, or cutting through a park. Imagine Trinity College's main entrance with a bridge over the Stein, visualise the Camac rushing under Heuston Station, or think of a mill (Dublin had dozens) or brewery behind any streetfront. Although most Dublin waters today look modest, some were more active in earlier times – and until recently, even a small watercourse greatly aided settlement. Rivers and streams, lakes and wetlands, all are fine sources of biodiversity too – Eanna ní Lamhna's "Wild Dublin" and Christopher Moriarty's "Down the Dodder" are good reads for wildlife context. The contribution made by watercourses was slow to be fully recognised, but now real effort is exerted to make them part of new developments, and culverting, once enthusiastically promoted, is generally discouraged; that said, we can still protect better against casual "taming" and overuse of hard engineering approaches such as canalisation. There are even plans for "daylighting" of at least one river section in the city.

Many of Dublin's waters, especially within the city and the more developed parts of the hinterland, are nowadays hidden for much of their lengths - piped, and in (thankfully rare) cases even "helping the drains" (but usually only surface water or stormwater lines). But they are not gone - and sometimes reminders of their existence are dramatic, as many a floodplain resident and business owner can testify, especially after the boom years, with so much land built on, and many front gardens paved over for parking. So, while Dublin – which looks full of streams on the map at the back – suffers severe flooding rarely – with parts sheltered against the kind of flood expected once in 10,000 years – when heavy rains do come, there can be nasty surprises. Notable offenders include the Tolka and the Dodder, and several of the latter's busy tributaries, as well as the Poddle and Camac, and to the north the Mayne, and the Santry in a few places, while Rochestown Stream elements (mostly buried), can rise up in Monkstown and Old Dunleary. And while the Liffey is of less concern near the city centre (though flooding in the Campshires *is* a risk, and Victoria and Wolfe Tone Quays have proven vulnerable in recent times), up in the Dublin Mountains it and its many tributaries can be very active. The October 2011 floods caused tragedy and chaos – but also showed how variable the effects of weather can be: the Dodder system (even modest parts such as the Killinarden and Jobstown Streams), and Poddle and Camac, played havoc, while the Tolka (helped by OPW and City Council work) and most northern streams, were mild in their effects (the Wad, Santry and Mayne Rivers did do "spot" damage out of proportion to their scale).

Most the bigger rivers and streams, and many smaller, played material roles in the histories of the areas through which they pass (many settlements lie on their banks, and industrial development can often be traced to watercourse lines, especially those of the Liffey, Dodder, Camac and Poddle), and some (at least the Liffey, Poddle, Camac, Stein and Bradogue) helped shape the city itself. Plans over the last century have also considered further use of rivers and streams, from the idea of major valleys as radial "green belts" to the formation of water features and other "breathing spaces" for new housing and business estates.

Anyway, I hope that this little volume intrigues others curious about Dublin's waters. Maybe it can also sometimes be of use to know of waters nearby – for which I also recommend use of classic Ordnance Survey maps, with their handy "Liable to Flooding" tags. Basic questions on sources are handled in end notes and the partial bibliography but for more on this, or for deeper detail, and queries about rights or educational use (requests from schools and colleges, Scout, Guide and other youth groups, community and historical associations, etc., are especially welcome), as well as plans, please mail *rer@raheny.com*. Corrections, and new information, or news of useful tools or sources, are all also always *very* welcome.

Rivers and Streams of Northern Dublin

- We begin our encounters with the rivers of Dublin at the boundaries with County Meath and the Irish Sea – we meet first the **River Delvin** (*An Ailbhine*), which partly contains the county boundary. Rising in the northern Garristown area, in the northwestern "corner" of Co. Dublin, with parts of its channel canalised and otherwise modified, it runs roughly northeast, for a total of over 18 kilometres (tributaries add perhaps 10 km more of channel to the system). Forming from branches from townlands including Princetown, Piercetown and Ashpark, and a longer one from Ardcath, the Delvin initially flows some way south of the county line. **Bartramstown River**, from near the Fourknocks archaeological site (just inside Co. Meath),

and with two small tributaries, joins, then **Garristown Stream** from one end of the village. After receiving three more southern tributaries the Delvin passes just north of Naul (see left), where cliffs of up to about 20m downstream gave the village its name; in these reaches the river is sometimes *the Roche*. Naul had at least two watermills in the past. Today there is a natural waterfall of over 5 metres, an artificial pond, and then a man-made cascade with a private hydro-electric plant. The river turns north and northeast, and more streams merge in, including one from east of Fourknocks. At Reynoldstown there is a multi-branched stream from the hills south of Naul. This latter has its primary source in two streams from Rath Little townland, which

join and pass, at Hynestown, through an early 20th century private reservoir, later picking up several small lines, and a bigger stream from the barrow site at Kitchenstown. In the next stretch (shown to right) there was once a flour mill at Oldmill Bridge. North of this, a small stream joins from Forty Acres. Then, just south of Stadalt, Saddlestown Stream flows in. Coming east via Saddlestown and Clinstown, and flowing wholly within Meath, with a branch from

Herbertstown to the south, this, probably the Delvin's largest tributary system, is said to have powered a mill too.

The Delvin forms the county boundary south of Stamullen (Stamullin), and runs under the M1 motorway. Within the former Gormanstown Demesne, long home to a Franciscan post-primary boys' school, it receives a small northern tributary where a millrace once drove a corn mill. Then a more active northern tributary from Mullaghteelin townland, historically mapped as Silver Stream, merges in, having passed Silverstream House (in recent years a priory of a Benedictine group) and having taken a tributary from Balloy. The Delvin then bends sharply south, and as it comes towards Gormanston Bridge, south of Gormanston itself, it takes in a stream (see left) from beyond Tobersool (long the site of a holy well). It then crosses the old northern road and meets a longer tributary, coming north past Balgaddy, from sources at Winnings and Whitestown.

A small southern tributary joins, the river passes a weir and becomes tidal (see left), and then it takes in one last stream from south and southwest, by the railway (photo on left below). Just after (shown right below), it reaches the tidal flats and the sea at Knocknagin, with megalithic remains some distance to either side of the mouth, north of Bremore (where there are also traces

of a small early harbour, and a tiny stream passing the site of the ruins of Lowther Lodge). The Delvin was sometimes known as the *Elvene*, *Delvyn*, *River Elvin* or *Elvin Water*.

- The **Hurley River** (*An Camán*, as in the hurling stick), rises in three branches in the southern part of Garristown's hilly hinterland, the central and primary one coming all the way from near the R130 road to Garristown, halfway between the townlands of Tobergregan and Nutstown. The branches converge in Charstown and the river flows west. A stream from a well by Tobergregan House merges with a long line coming south from near Primatestown,

and they join the Hurley in western Charstown. The line from Primatestown is shown on some maps as actually meeting a later stretch of the Hurley, which would be curious but may represent some form of artificial cross-connection. At any rate, the river then runs southwest, and west, across the N3 (Dublin-Navan) road, coming to Meath (see left), and crossing the townland of Knavinstown. In these reaches a number of field drains join. The Hurley passes through the northern part of the Tayto (soon to be Emerald) Park attraction, crossed by a bridge on the newer access road, and continues northwest and then north. In this stretch, the river passes Macetown with its castle ruins and remnants of a mill complex, including

part of a millrace. Later it flows to the west of Rathfeigh, with a historic mound and, to the north, another former mill complex. The Hurley turns to run east and northeast, meeting a tributary at Timoole. It then goes north to pass just east of Athcarne Castle and carry on to its dual confluence, the main channel by a road crossing at the edge of the former Annesbrook Demesne, and a secondary channel just downstream, with the Nanny Water or River Nanny, which it nearly matches in volume. The Nanny goes north to pass Duleek and then swings east to run under the M1, and later to go by Julianstown and Ballygarth; it has its mouth at the southern end of Laytown.

- **Bremore River,** a modest flow in northern Balbriggan, heading roughly east from a spring in the western Clonard area, in a field northeast of the M1. A tributary forms from field drains towards Flemingtown, and a possible link to the old Lady Well there, and runs south, meeting

a stream from south of Martello Terrace and paralleling Moylaragh Gardens. A branch heads off to the east parallel to Moylaragh Road, and the line then joins with the stream from Clonard, the Bremore continuing south of Moylaragh Crescent and Park. The river goes on to meet Drogheda Street, where it takes in the branch mentioned above and the historic flow it collected from around Bremore Castle (under reconstruction for Fingal Council for some years) and adjacent church ruins. The Bremore then passes south of St Molaga's School. The little river, most of its system buried but causing road flooding on occasion, comes into the open for the last time east of the railway line, and winds to the rocky shore (see left) and across the mudflats just northwest of Balbriggan's rather ruinous Martello tower at the Black Rocks.

- Cutting through Balbriggan town centre is the **Bracken River** or *Matt River* (*Abhainn Mhata*) (in one source, the **Glebe North Stream**), which once formed a mill pond (driving at least one major factory in the town). The river has two source sub-systems, the Matt and what is sometimes called the Inch.

The western sub-system, the Matt River (or Stream), comes from two branches. One, the **Walshestown Stream** at times, runs southeast from rising ground north of Walshestown, goes east and then passes under the M1 just south of the interchange at Rowans Little. It meets, at Courtlough, two small southern lines from towards Hedgestown, and turns north, running close to the M1 for much of its length. South of Decoy Bridge, it meets its other branch, from the Bog of the Ring and beyond (west as far as Kitchenstown) – with in turn several tributaries, including from Tougher, Belgee and Balrickard. Additional streamlets, such as from a spring at Turkinstown, flow in, before the river goes under Matt Bridge. It then takes in one more western tributary before turning to meet the eastern line.

The eastern sub-system, sometimes *Inch Stream*, sometimes *Salmon Stream*, rises between Salmon and Cross of the Cage, runs west and passes under Dennis Bridge, then meets what seems to be a small flow from the south. The Inch is linked to Knock Lake, a shallow artificial water body and former reservoir (and the largest lake in Fingal) between

Knock and Bowhill; this lake was made in the 1820s, acquired in 1974 by pipe-maker Wavin, and in 1993, rights over it were donated to a local angling association.

The stream passes just west of the village of Balrothery, near a holy well. At Gardeners-Hill, just south of Balbriggan, it takes in the little **Tanner's Water** from south and a little east, formed from branches east of Balrothery, west of Ardgillan Demense and north of Strifeland.

The Inch goes on to meet the western line of the Bracken system in Stephenstown, in a damp area where the remains of a mill complex, including a kilometre-long southern millrace (formerly driving a corn mill), can still be seen. The unified Bracken / Matt flows north (see left), by an area of light industrial and business premises. It traverses a green area parallel to Dublin Road, where another millrace used to run in parallel, and another corn mill existed; further *downstream*, in a former marshy area, was a mill pond, perhaps connected to the town's once active drapery industry.

Then, beyond the old bridge on the main road, the river bends northeast and falls gently through a small green area (see below) and

passes under the railway (picture to right) to come to the town's small harbour, where the estuary was reshaped many years ago to facilitate the making of the modest port. This area once contained a salt works.

- The former Hampton Demesne south of Balbriggan (long a key Hamilton family estate, later the site of an Irish language summer college) features multiple small stream lines and a pond, with the streams combining and passing under the railway line to flow to the sea.

- At Skerries is **Brook Stream** or **Mill Stream** (or just **The Brook**, or **Skerries Stream**; see right), an offtake from which feeds the millpond at Skerries Mills (visitable, with functioning watermill and windmills). The stream has at least three source lines: **Milverton Stream** (traces of an unusual watermill were found on the Milverton Estate) or possibly **Ardla Stream**, from between Strifeland and Black Hills, passing St Movee's Well. The Milverton merges with a second branch from towards Baltrasna, further north, while a third line comes from Killalane via Lough-barn, passing a holy well north of Balcunnin and receiving a tributary from towards Man of War (southwest). Below Skerries Mills there is a pond, with the remnants of a holy well, and the stream goes on to the sea just south of the village.

- Next is a streamlet running north and east from the Hacketstown area, reaching the sea at Holmpatrick, facing Shenick's Island. The origins of this stream once stretched further south than they now seem to, with a branch coming from west of the rail line.

- A stream sometimes called **Lane Stream** runs east from between Dellabrown and Ballykea, north by the hill at Popeshall and then northeast to come to the sea southeast of Holmpatrick. In one modern study this was labelled as the "Rush Road Stream."

- A small stream from nearby farmland at Thomastown northwest of Loughshinny village crosses the southern end of Loughshinny Beach (see above); a flow further north on the beach is artificial in origin, taking surface water from drains in the village area to the sea.

- South of Loughshinny, **St Catherine's Stream** (sometimes *Balcunnin Stream*) goes southeast from the townland of Balcunnin, passing Baldongan, going south and then southeast again. It passes the former St Catherine's Church and Holy Well, north of Rush Demesne and the site of Kenure House, and reaches the sea at the aptly named Brook's End (Cove). Just to the northeast, another small flow, probably natural, runs south to the sea.

- **Kenure Stream** runs east in the southern part of the former demesne lands north of Rush, with some water coming in from the north. Much culverted, as along Palmer Road, this small stream is seen in its final run to the sea, in a small valley south of the headland at the northern end of Rush's North Beach, and then crossing the beach itself (picture above).

- **Rush (Town) Stream** (or *Brook Stream*), which floods some of Rush's newer areas on occasion, flows from Tyrrelstown, just beyond the 15th railway milepost from Dublin, with several inflows, one coming south from Rush's former demesne. The stream goes east, then northeast, skirting the town's core, to reach the sea by crossing North Beach (photo to right).
- A small shallow stream flows onto Rush's South Beach within Rogerstown Estuary.
- A stream from Rathartan, west of Rush, flows roughly east, then southwest, reaching the sea near the end of Rogerstown Estuary, where Channel Road meets Spout Road, and a ford across the whole estuary used to reach land. Its final stretch is piped, entering the estuary alongside a surface water drain.
- Just east of the stream from Rathartan, **Bride's Stream** (also maybe *Rathmooney Stream*) comes to the estuary, with twin lines to the coast (one of which was upgraded in recent times).

The stream (etymology unknown but *Bride* in such contexts usually refers to Ireland's second patron, St Brigid) runs from Ballymaguire near Oberstown and the old workhouse. It goes on via the townland of Causestown and the former locale of a hamlet, Raheny, now seeing major housing development, just east of Lusk (from where it takes a streamlet tributary). At Whitestown, it merges with **Jone's Stream** (*Palmerstown* or *Collinstown Stream* are also seen). This latter runs from the townland of Palmerstown, passes the site of Caddell's Well, flows under Collinstown Bridge, and takes in a short tributary at Tyrrelstown. Between two confluence points there is a millrace passing a former corn mill (see above left), opposite Spout Road. The unified Bride flows on, past the site of St Maurus's or St Maur's Well, to Rogerstown Estuary, splitting for its final run as mentioned.

The eastern branch or distributary takes in a short flow from a nearby well or spring before running to the coast. The principal, western, mouth is shown right of centre below.

Documentation for the ESB Interconnector project (the cable of which makes landfall near Rush) labels Bride's Stream as "Whitestown Stream," apparently based on location.
Each of Bride's and Jone's Streams is a little over 3.5 km long, to their meeting, and one of them may also sometimes be known as the **Lusk River**.
The latter two "late" Rogerstown streams join in tidal lands just offshore.

- A small stream coming to the sea between the ends of Channel Road and Rogerstown Lane, on a shingly stretch of coast. This stream, which takes in field drains, rises by the road from Lusk to Rush, between former well sites, a little northwest of the joint Rush and Lusk railway station, and crosses under the Dublin-Belfast railway line.
A little further west of the above, a flow that looks like a small stream coming to sea where the railway line meets the coast, to the east of the Balleally's rubbish dump, appears to be a field drain, perhaps connected to the source of the previously mentioned stream. The next stretch of coast previously featured other short drainage lines in and by the inter-tidal area.

- **Balleally Stream** (possibly also **Regles Stream**), rising northwest of Lusk village, within the townland of Regles, and going southeast to pass near the round tower at St MacCullin's Church. Within Lusk, it runs partly through culverts, but can be seen (albeit with difficulty) from Barrack Lane to the green space at Chapel Farm, and (with less difficulty), part-encircling the civic space there, which was fairly recently just a field. Around this space, several small tributaries used to join, at least two from local wells or springs, and at least one still appears to do so. The stream goes southwest past the site of St MacCullin's Well, then turns to run south through the western Lusk area. Continuing south, and passing another holy well site at Bridetree, the stream tends southwest, flows via Balleally West, and then turns to the south-east, to pick up a small stream from near Newhaggard House and come to the sea at Newhaggard townland, directly south of Lusk, west of the Balleally tiphead.

- **Ballyboughal** (**Ballyboghil**) **River**, the main flow to Rogerstown Estuary and 5th of the county, rises northeast of Tobergregan in the hills just south of Garristown, and runs more than 15 km to its estuary, itself 4.5-5 km long. It flows east, passing Adamstown, then turns southeast at Grallagh, bending near Grallagh Cemetery, and going via Wyanstown and Leastown, where a stream comes in from Nutstown and Wyestown via Brownscross.

The first major tributary is **Daws River** (see right, sometimes known as **Clonmethan Stream**), from west of Jordanstown (with drainage connections back to Rath just inside Meath). It flows by Clonmethan and turns northeast at Oldtown, joining the Ballyboghil at Westpalstown.
Just after, still in Westpalstown, another short stream, runs in, having come from south of

Oldtown. The Ballyboghil then flows to (see below left), and through (below right, with guage), Ballyboughal itself, after which it receives a small stream from Drishoge and Belinstown. Next, **Richardstown River** (maybe **Grallagh Stream** at times), from Grallagh, Brownstown and other townlands, joins. Just to the west of the inflow of this river the great school and nunnery of Grace Dieu and its holy well lay, and a later millrace (some traces remain). *See below for a note on a link, which has existed for more than 120 years, from this section of the Ballyboghil River to the Turvey River to the south.* Near the sea, below Daws (or Maid's) Bridge, and south of Coldwinters, a small flow from Thomondtown joins, then the leading tributary, the Corduff River, meanders in.

Corduff River *(or **Stream**, upstream sometimes **Ballough Stream** or **River**, historically also the **Nine-Stream River**)* forms from many branches east and west of the M1 motorway (from locales including Hedgestown, Kitchenstown and Beldaragh). One branch comes from well to the west, near Nags Head, southeast of Naul, flowing via Clonany and Joinery Bridges. Parts of the Corduff system are seen at the twin service stations either side of the M1 motorway near Lusk (one such stream shown to right).

Later tributaries include **Wimbletown Stream,** from three branches from the hills south of Naul, north of Damastown, with a stream from Brownstown, and a parallel branch from Bettyville via Wimbletown Bridge; these combine just west of the M1 motorway. The Corduff runs south, at times parallel with the M1, through Coldwinters, crosses the road to Lusk a little northeast of Blake's Cross (see weedy stretch to the left) and tends southeast. The river joins the Ballyboghil in the latter's winding tidal reaches and the merged flow broadens, passing under the railway line, and proceeds to the open sea.

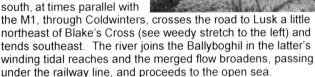

Several streams mentioned earlier run into the combined estuary.
- A stream at Beaverstown (no name traced), now flowing largely within a golf course but originating further south, and then **Portrane (***Portraine***) Stream**, a modest flow from multiple channels on the old Commons of Portrane by the demesne, then running straight north across fields and on with a clear course across the mud flats. It isn't clear if the straight run is a natural watercourse or part of a field drain system (it certainly receives field drains) - a canal was made around here some centuries ago and this could be its modern form. At any rate, the water-course was the subject of an unusual deal in the 18th century when part of the Commons was allocated to a family in return for looking after the watercourse – descendants still live nearby.

- Small streams on Lambay Island (one historical source describes it as "well-watered with numerous streams"). No stream names are recorded but the outfall of the longest, from the western slopes of Knockbane, north of Tinian Hill, running east, then northeast, in the north-eastern part of the island, is reflected in Freshwater Bay's name. A second, very short, stream develops from a spring, and Raven's Well, north of Heath Hill, and flows to Seal Hole. A third flow goes from west of Raven's Rock, near Trinity Well, to Carnoon Bay; some sources suggest that this was also fed from Trinity Well, while others appear to show a flow from the well to the west, which sinks soon after, and perhaps resurfaces near the curtilage of Lambay's castle. A short fourth stream near the castle, running west just south of the chapel, was culverted in the 19th century. Stream areas on the island have yielded interesting archaeological finds.

- Southwest of Donabate, in Corballis and southern Balcarrick, there were historically linked stream lines running east and south, and the latter still exists, mostly within golf course lands, coming to the sea at the eastern end of the Broadmeadow Estuary, inland of and near the beginning of the peninsula that runs south and holds a well-known golf course, The Island.

- **Turvey River** (or **Stream**, possibly also **Belinstown Stream**, or occasionally **Donabate River**), rises in Baldurgan townland, and passes through Cookstown (which holds a holy well) and Brownstown, where it takes in a short tributary from Magillstown. It then runs on towards both the old and new (M1) northern roads. The Turvey comes  east to Newbridge Regional Park (the former Cobbe estate / Newbridge Demesne) from the west, flowing in beside a side entrance. Joined by small flows and field drains, it becomes known as the **River Pill** (from an old word for an estuarine area, also met in central Dublin at the outfall of the Bradogue) as it crosses the park (see one bridge above and the ongoing course to the left). The river then runs under Hearse Road and onwards to reach the coast in a remote stretch of coast south of Donabate. Just before its current mouth (the sea used to come much further inland), the Turvey is joined by a stream from the west (passing Kilcrea). The river comes to the sea under  the railway embankment, away from public paths, and at low tide carries on over the mud.

Note: Just upstream of the M1 an open channel, existing for over 120 years (at least), links a fragmented section of the lower Ballyboghil to the Turvey, and in times of flooding can carry large volumes of water south (pictures suggest that the flow can at times be rather more than that of the Turvey proper to this point and reports suggest up to double volume); this channel *may* be known as the *Middle Stream*.

- **Lissenhall Stream** (sometimes **Staffordstown Stream**), from eastern and western branches meeting in southern Belinstown. It flows roughly south via Sunday Well Bridge (named for a nearby well, the waters of which once joined the stream, and may still do so discreetly) and then running east and southeast. Towards the coast it receives what appear to be field drains, and a channel flowing south and west from another well site, before curving southeast and then turning sharply north, and then again east, making a final run to sea.

See left for a view of the well-maintained mouth of the Lissenhall immediately south of Newport House.
- The **Broadmeadow River** (*Broad Meadow Water*, earlier *Gower/Gowre Water*, derived from the Irish form, *Abhainn Ghabhra*), with the 4th greatest flow in the traditional County Dublin (3rd if excluding tributaries), rises near Dunshaughlin in County Meath. Its main line is about 25 km long, with tributary lines totalling circa 60 km more - and a

long estuary (around 5 km) from western Swords to Malahide. The Broadmeadow forms from two main flows. One, sometimes **Ratoath Stream**, comes from branches east of Dunshaughlin, the most northerly of which, from by the site of St Shaughlin's Well, crosses Grangend Common. This stream later passes discreetly through Ratoath village itself (see right).
The other main stream, longer but often noted as secondary, and sometimes *Dunshaughlin Stream*, runs from Garretstown, on higher ground northeast of Dunshaughlin village.

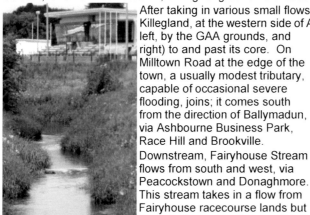

After taking in various small flows, the branches meet at Killegland, at the western side of Ashbourne, and pass (shown left, by the GAA grounds, and right) to and past its core. On Milltown Road at the edge of the town, a usually modest tributary, capable of occasional severe flooding, joins; it comes south from the direction of Ballymadun, via Ashbourne Business Park, Race Hill and Brookville. Downstream, Fairyhouse Stream flows from south and west, via Peacockstown and Donaghmore. This stream takes in a flow from Fairyhouse racecourse lands but

its main line comes from well to the west, between Powder Lough and Wilkinstown; it meets other tributaries and field drains, goes under Donaghmore Bridge and merges into the Broadmeadow northwest of Greenoge. Another modest southern tributary follows at Greenoge, its main line coming from beyond Wotton Bridge to the southwest. By the county boundary, marking it for a stretch, the **Dun Water**, with several branches - including from Thornton (Thorntown) - flows in from the south, having passed Dun Bridge, between two small northern flows (possibly field drains).
In its middle course (see left), the river meets a stream just above Rowlestown Bridge, from Killsallaghan via Killossery; just above

the meeting point there used to be a cornmill north of the churchyard. More streams are found at Lispopple (flowing south and east, from a crossroads historically called Wren's Nest, on the road to Oldtown) and Warblestown (one going south, one north from branches beyond Roganstown House, one of which comes in turn, via a golf course, from around Lubbers or Lovers Wood), and either side of Saucerstown (one is sometimes **Saucerstown Stream**; the lower comes from past Rathbeal).

There is also a curious diversion from the river in this vicinity, rejoining near Ashton Green, having met field drainage (the line seems clear but operation could not be fully checked), and then small flows around Balheary (from Oldtown and Clonmethan).

The Broadmeadow then arrives to Lissenhall Bridge (pictured below)

and then the N1 road (and another bridge, as pictured below).

Just east of the N1, northeast of Swords, the **Ward River** (to the Anglo-Normans, *La Garde*, *Abhainn Bharda* in Irish) joins the Broadmeadow, as its principal, and final real tributary. The Ward (also **Swords River**, especially near the town) rises in Meath, by Killester townland, north of Folistown, west of Rathbeggan and south of Fairyhouse Racecourse (just north of one of the Tolka's Pinkeen Stream tributaries), and flows over 18 kilometres to meet the Broadmeadow.

The river runs southeast through Boolies to Nuttstown, where it swings to an easterly line, goes past Kilbride, then curves east at Irishtown. Early tributaries include one from the southern edge of Fairyhouse Racecourse, one at Nuttstown, and one at Baytown.

Mabestown Stream is next, joining within Irishtown townland, then a sizeable flow coming north and northeast from between Gallanstown and Yellow Walls to merge in at Ward Lower, just east of the M2 - and a small flow at Coolquoy. Downstream of Coolatrath Bridge, a stream joins at Coollatrath East, south of Corrstown, having formed from branches at Loughlinstown and Ballyhack in Meath. **Shallon Stream** (shown right) joins around Chapelmidway, coming via Bishopswood from beyond Cherryhound towards Hollywoodrath, with a branch from Kilshane. The next tributary is **St Margaret's Stream**, from the Hunstown area north of Finglas and the M50 motorway, west of Coldwinters. Branches form near Huntstown Power Plant and a quarry, and to the west. The stream passes Kilshane Cross and under Kilshane Bridge, and near holy well sites (at Kilshane and St Margaret's) before and after former Plunkett stronghold Dunsoghly Castle, and also passes Kilreesk Bridge. It takes tributaries in turn from Pass-if-you-can (near the Harristown source of the Santry) and between the townland of Forrest and Barberstown House. The stream joins the Ward at Skephubble, at the edge of a golf course.

A stream runs south via Laurestown and Toberburr to merge in at Westereave, and shortly after the Ward gains tree cover, that broadens into full woodland after it passes Knocksedan Bridge, south of a former archaeological enclosed site. In this valley, once overlooked by Brazil House from the north, a stream came southeast, and used to feature a pond in the Brazil area; it joined near where Brackenstown House stood by the river's southern bank. Just north of the house a pond or reservoir was formed (outflow below),

with an adjacent millrace which fed a flour mill, also repurposed as a sawmill. A smaller pond lay to the east, later infilled. South of that was a long narrow pond, once "Old Pond"; this shrank over time, becoming mostly marsh, with a smaller remnant pond formed at its eastern end.

The river passes under a bridge and continues towards Swords (see left). This stretch of the river is sometimes known locally as the "Jacko" and the one remaining pond sometimes as Usher's or Ussher's Lake; the river and pond provided a clothes-washing and bathing place into the 1970s.
The river valley has largely been reworked into the Ward River Valley Park (see below), sometimes locally called "Jacko Park".

Small flows also existed in the south of the park. There was once a windmill where the river used to wind below the townland of Windmilllands, and just to the east there is a remnant of a former millrace below Gallows Hill at the edge of Swords town centre – a flow that used to drive a cornmill.
Exiting the

main body of the park, the Ward turns to run north, to parallel the main street of Swords, down-hill from the round tower (since built into a later church) and near St Colmcille's (Columba's) Well. The well is linked to the naming of the town (*Sord Colmcille* in Irish, from a blessing of the well's waters in the 6th century), and was restored in 1991/1992 by Swords Historical Society and the local authority, although due to risk of pollution and damage, the well chamber (see left) has mostly been kept locked for many years. The round tower is the last remaining element of the monastic settlement reputedly founded by St Colmcille, the third of Ireland's patron saints, near the well and overlooking the Ward River.

A small stream from branches in Rathingle runs east, and, as it approaches central Swords, turns north, running partly on the surface. It passes the site of Slips Well, near the modern housing of Highfield Downs, turns east by Bell's Lane (see left) and northeast, and, going through the car park of the former pub, the Lord Mayor's (to right), joins as the Ward begins its northerly run.

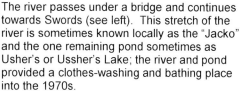

The Ward flows north, along one edge of the town's central park, passing Swords Castle (restored and opened to the public after impressive archaeological work). Facing the castle wall, it discreetly receives the flow of a stream which has come north from Gallows Hill via the now-hidden Brigean Bridge on Rathbeale Road, south of the site of Doyle's Well. The Ward goes on, past an industrial site, once on the edge of Swords but now well within its urban area.

A final tributary comes east from branches northeast and southeast of the site of Glasmore Abbey and St Cronan's Well. A little of the southerly branch can be searched out where Glasmore Park meets Rathbeale Road. The branches combine in culvert north of Swords Shopping Centre, and this stream of no known name flows along Watery Lane and into the Ward by Scotchstone Bridge (see left), to the west of the upper end of Main Street.

The Ward goes on north in a "green corridor" with, partway along, a curious mid-stream concrete structure (see above right, possibly a calming device).

The Ward runs on under Balheary Bridge (see right), almost merged with Lissenhall Bridge, as seen on page 11 (both bridges lay on the old northern road, a short line of which remains, crumbling gently, west of today's main road).

The Ward flows on under the new northern road and curves slightly northeast to its confluence with the Broadmeadow (below left), in an overgrown pocket of land on the margins of Swords, close to the beginning of the ca. 5 km-long Broadmeadow Estuary.

The Broadmeadow passes east through marshy land (parts are named, such as Old Marsh and Horse Marsh), and opens into the wide estuary, a little east of the over-flying M1 (see below). There is a main channel continuing the river line in the north, and a narrower channel looping to the south. Along the northern shore there is a permanent presence of swans.

The Broadmeadow Estuary is in turn split by the rail line (the Dublin to Belfast line) north of Malahide, which rises on an embankment, with a narrow gap crossed by a viaduct, and then comes down in the direction of Donabate.

The inner part of the estuary (see right) is now a semi-impounded tidal lagoon, while the outer, towards Malahide village itself, is fully tidal, and mostly mud at low tide. In the late 18th century, the then Lord Talbot de Malahide secured an act to allow him to build a canal from the head of

the estuary inland; it seems that construction even commenced, but was later abandoned.

At least five small watercourses enter the inner Broadmeadow estuary. To the south a short stream joins from the townland of Mantua (see left), then a piped longer one from the townland of Swords Demesne. Then from the north, the Lissenhall enters within the first kilometre, and then the Gaybrook Stream and a nearby small stream, both flowing north. The Turvey River flows into the outer estuary in Kilcrea.

The Broadmeadow Estuary passes to the sea by way of a gap of maybe 200 metres between Malahide proper and Malahide Point on Malahide Island. Aside from the well-known haunt of swans on the north shore beyond the motorway flyover, the estuary is rich with more than 60 species of bird.

- **Gaybrook Stream** has its mouth, within the inner Broadmeadow Estuary, in the Yellow Walls area west of Malahide village. This was once site of a distinct village which at its peak was larger than Malahide, but is now an integral part of the area.

The Gaybrook's main line comes from Fosterstown, within the northern reaches of the *Boroimhe* residential development off the Swords Road. Flowing underground, it crosses Swords Road and

enters *Airside* retail and business park, where it receives small inflows and passes the site of St Werburgh's Well in Crowscastle, near a water feature (probably not connected) and Ryanair's HQ and other

offices (see above). It continues east, past Feltrim Business Park and the M1 motorway, then runs through Drinan, tending southeast as it approaches the *Gainsborough* estate.

A second branch rises in southern Crowscastle, in the central part of the *Airside* development, east of the retail park (see right); this flows by two ponds (one shown to right), to and under the *Holywell* development, reemerging east of Ashdale Road, and joining the prime branch south of the western end of Gainsborough Park. This branch takes in a culverted flow from the site of Drynam House and an old well.

The Gaybrook flows on roughly north, passing between Talbot Avenue and Castle Downs (see below), and

travelling under the road from Malahide to Swords, then receiving another tributary between Seabury Court and Downs, from Mountgorry. It next runs under Barrack Bridge at Old Yellow Walls Road. After falling several metres over a short distance, the stream enters Caves Marsh, an interesting complex of land types in a small space, including both freshwater and that rare and important type – salt marsh – and flows north, broadening into a sub-estuary of the Broadmeadow Estuary (three photos below). There are paths through the marsh area.

On a point of industrial history, what is today a small stream – but then had greater flow and more tributaries – once drove one of the largest cotton mills in Ireland, a five-storey facility with a mill pond and an "Arkwright mill" system.

- Just to the east of the lower Gaybrook, a small stream also reaches the sea, coming out of culvert and cutting through the eastern edge of Caves Marsh (see right), and then turning sharply seawards (shown below). This modest flow rises from  now-buried sources to the southeast, at what is now the *Sonesta* development (where there were historically two springs or wells) and at Yellow Walls Road. Supplemented by a small channel within the marsh complex, it finally comes to the sea over a shingly shore (see right), near a local yacht club.

A little further to the east again, and inland, Malahide (Castle) Demesne, now a regional park, holds at least one spring, and wet channels and field drains, but no streams.

- The **Sluice River** (*Abhainn na Comhla*) forms from (north to south) the **Forrest Little**, **Wad** and **Kealy's Streams**, three of Dublin Airport's seven catchments. The Forrest Little Stream rises in Pickardstown, receives flows from Forrest Great and Forrest Little, crosses Fosters-town and heads for Kinsealy (Kinsaley). The Wad and Kealy's Streams rise within the airport campus, the Wad historically near the former TEAM-FLS building, Kealy's midway between Terminal 1's car park and the "lobby" of Terminal 2; a diversion from Kealy's Stream to the Wad was made near TEAM in the late 1990s. Both flow eastwards, and pass under the Swords Road, the Wad just north of the airport access roundabout (and then bending towards Stockhole Lane), and Kealy's Stream (to right) running to the long-term parking area entrance, and then across that area. Both then run under the M1 and re-emerge in parallel into the Baskin area (a section of the Wad shown to left), where the gap between them grows. The  Forrest Little and Wad Streams merge at a small artificial lake in the Greenwood part of the Abbeville (Abbeyville) estate, and their combined flow was used to form the two lakes by Abbeville House (Charlie Haughey's former home). South-east of Abbeville, Kealy's Stream turns sharply north and joins; it historically supplied a small brewery north of Baskin Lane.

The Sluice runs under Malahide Road at Kinsaley Bridge, north of the St Olave's housing estate, and on via eastern Kinsealy, picking up a streamlet. After crossing the old trotting track in the original Portmarnock village, it goes under the railway, then heads east, then roughly south, part-encircling the Sluice River Marsh, a rare feature for Dublin, containing a mix of freshwater and saltwater marshland. Early in this stage, shortly after passing the railway, it receives the Hazelbrook Stream.

Coming east and southeast from Marshallstown (south of Kettle's Lane) and with a streamlet joining from Feltrim, **Hazelbrook Stream** (shown left) takes in a tiny stream coming west, passing under the railway, from Beechwood, then flows in turn east under the railway, then south to the Sluice. As it passes around the marsh, the Sluice meets flows from west of Portmarnock, from what is today Malahide Golf Club's course. It then approaches (see above right), and passes under, Sluice or Portmarnock Bridge at the western edge of the modern Portmarnock village, to reach tidal lands (see left) at the head of Baldoyle Bay (which also holds the Mayne River's estuary to the south). Near the mouth, a secondary channel or field drain (visible at righthand side of photo) joins from the western edge of the marsh. Baldoyle Bay and its mudflats are popular with birds. The Sluice system floods occasionally in Streamstown and by the former trotting track in Old Portmarnock, and more rarely but potentially severely towards the coast.

- The **Mayne River** (or *Maine* or *Moyne River*) has its lead source in the **Cuckoo Stream**, Dublin Airport's prime drainage line (which once defined part of the then Collinstown Aerodrome's boundary), falling about 60 metres along its course. It comes from Dunbro beyond Collinstown, emerging from airport lands opposite the southern part of the ALSAA sports complex, then runs to Toberbunny between the old Swords Road and the M1 motorway, and passes ALSAA's pitch & putt course. It goes south of Spring Hill and Lime Hill, north of Balgriffin Cemetery, crosses Malahide Road at now-hidden St Doolagh's Bridge, and runs along the northern edge of Fingal Cemetery (where an old estate bridge, Sarah's Bridge, survives). Meeting a small northern flow, then one from the south (culverted except at the eastern edge of Fingal Cemetery) with its origin to the west in the Belcamp Estate lands, the Cuckoo goes southeast, running under Wellfield Bridge to meet the other major branch of the Mayne.

That major branch - **Turnapin Stream** - from upper Harristown, and with possible connections to the upper Santry River, and, it may be, to Kealy's Stream, runs east. It then meanders north into the airport campus for a stretch. It later passes between the RCSI (Royal College of Surgeons) secondary sports grounds at Dardistown and a karting track, and then under the old Swords Road at the no-longer-visible Turnapin Bridge, coming to eastern Dardistown. It meets a small tributary from further north at the back of a little industrial estate, and then another, draining southern airport lands, near the M50. The stream passes the M1-M50 interchange (it can be glimpsed from the northbound downward slip road), and Clonshaugh.

Moving under the road to run north of the N32, the stream goes through southern Belcamp lands, where it receives field drainage, takes a tributary from northern Priorswood, and forms ponds near the former Belcamp College school. It then exits Belcamp to go below Balgriffin village (see left), and crosses Balgriffin Park lands

(on right, with fence over mouth of small tributary) where it once fed ponds.

The Turnapin meets the Cuckoo just east of the Castlemoyne development, to finally form the Mayne River.

The river runs under the Hole in the Wall Road and crosses under the railway line via the historical "Red Arches" bridge (one arch shown left). A stream coming east from past Drumnigh - south of the Trinity Gaels GAA grounds - and under the railway north of Moyne Road, joins south of Moyne Park, and then Grange Stream runs in.

Grange Stream rises, buried, near Grangemore housing estate in western Donaghmede and runs east in culvert. It flows south of the church remains at Grange Abbey (a small local church; there was no abbey, though it belonged to the Priory of All Hallows on the edge of medieval Dublin), where it once fed fish ponds; the last pond was filled in with rubble from Grange House in 1972, and archaeologists later found a hoard of gold sovereigns from the early 1800s there. The Grange continues by Donaghmede Park, in which it formed a feature for some years (it is now piped), and comes to Baldoyle (the last culverting of the

stream was at Marian Park). In the 19th century, there were several connections in this area, near the rail-way, between this stream and the Daunagh Water, at least one of which was still mapped as partly open in recent times. Where Grange Road and Willie Nolan Road part, historically the site of Lark Hill Bridge, a stream joins from the former marsh remade as Seagrange Park (sometimes **Seagrange Park Stream**, see left), having followed the railway line for a stretch near Bayside DART Station. Sweeney also notes a watercourse coming in from Brookstone Road. The Seagrange Park flow is a cut-off remnant of the original Daunagh Water course; it was diverted in the 19th century to its outfall on the coast by today's Bayside.

Grange Stream then goes north across former Baldoyle Racecourse lands (see left), mostly in culvert, coming to the Mayne (shown to right). An artificial "relief line" takes water from the stream to the coast just south of Baldoyle's Roman Catholic church.

A small flow used to join the final stretch of the Grange Stream from the west. Rising at what is now the corner of Father Collins Park and the main street of the Clongriffin development, it ran north of Grange Park and passed under the railway to the south of the Red Arches - then merged in just east of Stapolin House (its line can be seen in the boundary between Stapolin Lawns and parts of Red Arches Road housing). The stream was still extant by Father Collins Park in the early 2000s, and even in the mid-2000s there was still a small stretch by the western side of the railway, and parts near Stapolin House, but by 2013, I could find no visible signs west of the railway; however, there is what

seems to be a remnant of its line (see right) within the former racecourse lands. The Mayne's final reaches flow more or less directly east (see left, from upstream).

The river reaches Baldoyle Bay at Mayne Bridge (view below right), with two exit channels and tidal backflow protection. There are areas of brackish marsh near its mouth.

The Mayne regularly floods at several locations; works have been carried out over decades to address this.

We now meet an area that was within the city until 1985, when Baldoyle, Sutton, Howth, and much of *Kilbarrack Lower* (mostly Bayside), were moved to the newly-defined Fingal area (a county-level authority from 1994); the city also gained some land. Starting with Grange Stream, I add gratefully to sources 1991's "The Rivers of Dublin" by City drainage surveyor Clair L. Sweeney (1923-1996; cf. biography page), also source for a 1978 Irish Times "Diary" piece. Mr. Sweeney's work and interest enabled him to share details on many waters, including the largely hidden, and even to outline courses for two possible

"lost streams" in the city centre. The book is a fine read for stream detail and for information about areas passed by streams and rivers; it also has a range of photographs, fine maps and a note on wells.

Streams of the Howth peninsula:
The hilly Howth peninsula features small (as the water runs off rapidly) clear streams and the Holy Well of St Fintan (in a private garden up from the cemeteries at Carrickbrack Road) - and at one time also had up to eight regular and two "petrifying" wells. Ireland's Eye off the coast has no streams but at least one spring (now mostly just a persistent damp patch).

Howth Head's watercourses include:
- A few small flows historically going to Claremont Strand, then four streams going north within the Howth Estate (also source for the Carrickbrack and Santa Sabina Streams flowing to the Sutton area; see further on in text), the courses of which have been modified over centuries.
One "Howth Estate stream" runs east of the Presbyterian manse, one by the Swan Pond (see right) and the volunteer-run National Transport Museum's main base, and one near Howth Castle ("captured" by the castle's drains).

Farthest east is the **Bloody Stream**, named for a reputed battle between invading Cambro-Normans and the Norse-Irish at Evora Bridge, a battle that the first Lord of Howth won (losing seven close relatives). It rises on the north slopes of Black Linn, Howth's highest peak, in the Ben of Howth area, runs through the estate's "Deer Park" golf facility and takes flows from near the castle (an outflow near here used to head for Black Jack's Well / Pond, near which it may still serve a moat; it then flowed, and may still flow, north to the treed area east of St Mary's Church before rejoining the main stream). West of the St Mary's Church lands, and downhill, there is a visible cascade (see left below) to below the main road. The

stream goes west along the northern (seaward) side of the road, buried deep, then cuts across the former Techcrete factory site – an idea of reopening the stream there as part of a major proposed redevelopment was raised in 2008 but 14 years on, has not progressed (nor has the housing planned). At least three of the Howth Estate streams meet at a "screen house" and go under the railway to come to the sea as one, in culvert, 415 metres west of Howth's West Pier.

- **Offington Stream**, a short watercourse from south of Balkill Park. Historical mapping shows it beginning by woodlands at what was then labelled *Balcil*; more recent maps suggest its head is today connected to drainage lines from further south and east. Priest's Well used to rise nearby but seems no longer visible. The stream runs north within the wooded boundary to the east of the Deer Park golf facility, meeting a small drainage line from the west early on. It was formerly just west of a winding part of the line of the Hill of Howth electric tramway (a rail circuit, taken out of service in 1959, linked to both the then Sutton & Baldoyle Station and Howth Station), crossing it just behind the former St Lawrence Hotel, where a walled path partly on the old tram alignment now rises steeply from the western end of Howth village (the tram approached Howth Station via a long-removed bridge over the Dublin Road). The Offington ran in a straight line under part of the St Lawrence Hotel into western Howth Harbour, then closer to the property line. The stream still passes under the development which replaced the hotel. Occasional flooding at a debris screen by the former hotel has been noted by the local authority.

- **Boggeen Stream** (or ***Boddeen Stream*** or ***Grey's/Gray's Brook***), from branches around Kitestown Road and near the Summit, the most remote rising in

the angle as Thormanby Road curves to a northerly line and running north along Thormanby Road, then sharply east, taking in some drainage lines, then northeast to form the main stream line. A streamlet from a well site west of Balkill Road (once Lighthouse Road) joins at the southern point of Thormanby Woods; another from a nearby spring flows north-west and northeast by the wood's edge. The Boggeen joins Grey's Lane (formerly Boggeen Lane, see left), and flows openly in a small valley (shown right) between the lower part of the lane and the old tram line, to the turn where Balglass Road comes to Main Street, where it goes into culvert; around here it probably takes in a small tributary coming from a spring to the west a few hundred metres uphill. It then goes west of the modern church, and passes "The Abbey" (St Mary's Church) - it is referenced in old property deeds - and, going west of the uphill site of Juan's Well and east of another well site, flows discreetly into the eastern part of the harbour.

- **Coulcour (Coolcour) Brook**, rising at a buried spring south of Thormanby Lodge, passing Casana View, and running along Thormanby Road, joined by a tiny flow at Dungriffin Road. The brook runs just east of the start of Cowbooter Lane (see left), from the site of Cannon Rock House, and then passes on in culvert to continue by the western side of the descending lane, running buried behind Brymore House Nursing Home and other buildings and emerging near the end of the lane for a short surface run. It then goes back into culvert, crosses Nashville Road and runs a little west of Kilrock Road, coming into the open in a deep valley to approach and pass Balscadden Road and fall to Balscadden Bay.

- **Whitewater Brook**, from the southern part of the Ben of Howth, under Black Linn, Howth's highest point, runs east, then south, passes Windgate Road, crosses Carrickbrack Road, then flows behind a public car park (see left) and crosses Thormanby Road. Meantime a stream emerges from a retaining wall at Carrickbrack Road, passes through ponds (see right) between

Carrickbrack and Thormanby Roads, and joins the Brook. The stream flows on under the Cliff Walk and tumbles to White-water Brook Cove, south of Gaskin's Leap and north of the Baily peninsula. The Brook flows mostly in the open.

- Three streamlets, from Thormanby Road to the sea east of Lion's Head, and from Windgate Road and Carrickbrack Road to Doldrum Bay (site of "The Needles" rocks).

- Two streams outfalling west of Drumleck Point, the larger **Balsaggart Stream**, the biggest watercourse by volume on Howth. Flowing at a grade of 1:9 from "the Cumulus" between the Ben of Howth and Shelmartin (on Howth Golf Club lands) to fall to the sea at Sheep Hole, the Balsaggart takes in a streamlet from Black Linn, passes Carrickbrack crag with its cave, and goes under Carrickbrack Road a little to the west of Ceanchor Road and the sites of two wells, one Balsaggart Well. Passing the Cliff Walk very near the coast (see right), it falls to the sea by Sheep Hole. The Balsaggart main course is largely above ground, and in times of flooding, part of its flow can cascade on to Carrickbrack Road.

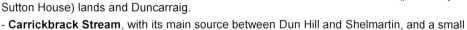

- Three small flows in coastal Sutton, between the former Sutton Castle Hotel (previously Sutton House) lands and Duncarraig.
- **Carrickbrack Stream**, with its main source between Dun Hill and Shelmartin, and a small

inflow near, crosses woods by Muck Rock, after which another flow joins. Exiting Howth Estate at Carrickbrack Road, near St Fintan's National School, it is joined by its main tributary, from above the old St Fintan's cemetery, which has a branch from the rocks north of the clubhouse of Howth Golf Club. It then follows the road downhill, to its mouth near where Strand and Carrickbrack Roads meet; the final stretch runs underground near a coastal monument and comes to Sutton Strand in a large-bore pipe (see above left).

- **Santa Sabina Stream**, rising south of the main avenue of Howth Demesne, flows almost directly west, south of the "Round Plantation" and in culvert to the Offington housing estate. Crossing Offington, it picks up small inflows and comes to Greenfield Road. It then traverses the Santa Sabina convent grounds, taking in an altered branch which now runs west of the convent, and, crossing to the small beach, has its primary mouth, with safety railings, 155 metres beyond that of the Carrickbrack Stream (see below; the photo on the right shows how the outfall can look after a storm, the stream manifesting as merely a line of damp sand).

The stretch of coast where the above two streams have their mouths, part of Sutton Strand, has three outfalls that flow even in a dry summer as well as several smaller (surface water) outlets. All of the outlets are often wholly or partly covered by sand deposits, which can reach back up the pipes. The southernmost of the three major outfalls carries the Carrickbrack, while the northernmost, with a retaining wall and some railings, seems to be the primary outfall of the Santa Sabina. The source of the middle one (see right) is most likely the original line of the Santa Sabina, possibly supplemented by a well or spring.

- **Kilbarrack Stream** (aka the **Daunagh Water**), today culverted right to the coast but still fully visible in the 1960s, rises east of the Santry Bypass, south of Turnapin, passes Priorswood and Darndale and crosses Malahide Road near Newtown Coolock. It runs north of, and angles towards, Tonlegee Road, traversing part of Ayrfield. At one time, it next ran openly in the parkland parallel to Millbrook Avenue, south of Foxhill and then just south of the Donahies Community School, crossing Streamville Road and flowing by the end of Laragh Close and north of *The Beeches* (before the current housing development, it featured in the garden of the house of that name when it was home to retail entrepreneur "Hector Grey"), then crossing Grange Road and paralleling St Donagh's Road. The part from Donaghmede to western Kilbarrack was culverted in the 1970s (the area around the culvert line is laid out as a linear park), that from Millbrook Road to Streamville Road in the early 1980s, and the final stretch to Grange Road in the late 1980s. 145 metres east of Grange Road, the stream takes in the waters of the buried holy well of St Donagh, which once had a 5-metre pool. About 400 metres downstream from Grange Road, near where it used to take in a small western tributary, the stream is divided in its culvert. The main course runs under Howth Junction DART Station, just after the parting of the Belfast and Howth lines, taking in water from two northern lines which used to connect to Grange Stream (their current status is unclear, although one appears to still run under Carndonagh and part of Baldoyle's industrial estate), today finally reaching the sea near Bayside Boulevard South, west of the old Kilbarrack church (Dublin's mariners' church once, supported by its own harbour levy). The other line, a modern diversion, runs roughly parallel to Kilbarrack Road, outfalling east of a local authority pumping station on the coastal embankment.

The Daunagh Water used to turn away from the coast (under where Bayside Crescent is today) and, gathering field drains, flowed northeast towards Baldoyle, crossed back under the railway branch to Howth, took in flows from west (linking back to one of the previously mentioned Grange Stream connections) and east, travelled in two branches through the marshy area then situated where Seagrange Park is now, and received Grange Stream and a small eastern flow just north of Lark Hill Bridge, finally joining the Mayne River (see Mayne River section above for detail from this point on). Diverted in the 19th century, it left a partial remnant, still visible at the eastern edge of Seagrange Park and along a stretch of DART line just beyond park bounds. Already in the 19th century, there was also some form of water-course link to Kilbarrack Road (then called *Killbarrack Lane*).

Rivers & Streams of Raheny

Raheny has four watercourses in fairly close order, two of which run wholly within the area. There are also several persistent "damp patches" (believed to be from active springs) sending water down the slopes behind Howth Road and Maywood / Bettyglen / Blackbanks to the coast road (James Larkin Road).

- **Blackbanks Stream**, today fully culverted, rises just south of Tonlegee Road and runs southwest, swinging east by Edenmore's St Monica's church, then going via a limestone area, once holding quarries and ponds, partly infilled as a landfill site in the 20th century. The tip was remade as Edenmore Park and the rest of the area developed as Woodbine. The stream goes on south of Grange-Woodbine Hall and by St Michael's House Special School (it runs in a "corridor" just uphill of the school buildings), in the shadow of Belmont Hill (or "Mount Olive" – once site of a windmill and now of a friary, housing and a hospice). Joined by two surface drainage channels, it also takes a full diversion of the upper Fox Stream, which drains the other side of the steep hill, and includes run-off from the modern Ashcroft and Tuscany Downs estates. The stream then goes under the railway and on between the Roseglen and Foxfield estates, it turns south at Greendale Road; its outfall is where Howth and James Larkin Roads meet (above right). Steps allow access to its mouth; the line runs on through tidal "lagoon" mud towards the main channel from the Santry.

- **Fox Stream** (or *Fox's Lane Watercourse*) used to come from the same limestone area north of Raheny village centre as the larger Blackbanks Stream, though from a nearer point, and its line lies on the opposite side of Belmont Hill, by Walmer House (now part of St Francis Hospice). However, as noted above, the upper course was cut over to the Blackbanks Stream, as it was thought that the Fox's railway culvert would not be able to handle increased water volumes when the former St Vincent's GAA facility, by

Raheny's old trotting track, was redeveloped for housing. From its railway under-pass it still flows, entirely in culvert, through the Foxfield housing estate, across Howth Road, and along Fox's Lane, to come to the coast (see bottom left, previous page) under James Larkin Road, just over 400 metres from Blackbanks Stream (with a surface water drain pipe passing near to outfall farther out into the "lagoon"). There was for a time a hamlet near the lane and stream mouth, "Raheny-by-the-Sea," of which one thatched cottage remains on Fox's Lane. This small stream once marked the northern bounds of Dublin City ("a little brook which is the end of the Liberties of the City of Dublin 130 perches north of the Mill of Raheny").

- The **Santry River** *(Abhainn Sheantraibh)*, from branches in Harristown and Dubber, and with small inflows from airport lands, travels about 11 km to the coast (and at lower tides, about 3 km more across the "lagoon" behind Bull Island to Sutton Creek and the open sea). The first Santry roots can be found near a lane in the former Harristown demesne, and it runs south of Dublin Airport, passing Harristown bus depot and taking in the Dubber branch; it then goes via Sillogue, visible within the municipal golf course, and northern Ballymun; the main channel of this early part is sometimes **Quinn's River**. Taking in a tiny northern branch, it comes to the former Santry Demesne; the part of the estate around the river and the former pleasure pond has been remade as a public park, while much of the remainder has been developed with housing, hotels, offices and shops. The Santry then passes under the old airport road. Both the park and the old airport road are subject to occasional flooding. Having run under the M1, the Santry features in Coolock's central valley, taking a diversion

from the Naniken (details below) at playing fields by Clonshaugh Industrial Estate. It runs through a linear park (see right & left), and forms a pond by the Stardust

Memorial Park. It flows past the Cadbury's factory and Irish base, and a tumulus on factory lands, before going under Malahide Road. There was until recent times a holy well on its banks near the site of St Brendan's Church (now holding the Church of St John the Evangelist); the well site is marked by a tree (see left) and the waters are drained discreetly into the Santry. The river runs on between Edenmore and Harmonstown, and continues, in a winding stretch (shown to right) including passage under the railway. It then crosses central Raheny, below the lands of the prominent Roman Catholic church

(Our Lady Mother of Divine Grace), where water from the buried Holy Well of St Assam joins and the river valley passes downhill from a bank building (shown left). It goes under a bridge on Howth Road (fitted with lighting and viewing points in 2004) in the modern core of Raheny, and on past the Scout Den and an old police station site, and via a bridge on the old Main Street. In the late reaches of the Santry, flooding occurs in some periods of heavy rain. The river flows on between the linked housing estates of Avondale and Maywood, and Manor House School, and comes to the former Bettyglen Estate, late of the Jamesons of whiskey fame, where it is piped for a

little, before winding through a small wooded area and entering a final culvert to the coast. It emerges alongside a surface water drain (above left), just over 600 m from the Fox Stream. It flows on across saltmarsh (see above right) towards Sutton Creek across the eastern part of the "lagoon" inshore of Bull Island. The river once had a watermill, "The Mill of Raheny," just north of Watermill Road, east of Millbrook Cottage – it was in ruins by the late 18th century – and a stone quay at its mouth. Flowing in the open except for short sections, this watercourse is occasionally named the **Raheny River**, as by Lord Ardilaun when he required that its flow, and that of the Naniken, not be disturbed by the construction of the Dublin-Howth tram line - and on one early map may be named **Skillings Glas** ("Glas" from the Irish for "a small stream") – but it seems increasingly likely that this was simply a placename. The coastal stretch including the Santry and Naniken mouths was called *Raheny Strand*, and featured on a 19th century postcard which referenced its oysters.

- The **Naniken River** (also found as Nanekin, Nannikin and Nanikin; An Nainicín in Irish) rises in culvert south of Santry Avenue (earlier Santry Lane, running from Santry village towards modern Ballymun), by Shanliss Way. It runs under the Santry Hall light industrial and business estate, and then under the old Swords Road. It passes below Schoolhouse Lane and the Oak Park streets, before going, via syphon, under the M1 motorway, just north of the Port Tunnel entrance. It then crosses the City Council-owned "Oscar Traynor lands", and in this vicinity much of its upper flow is diverted since the 1960s to the Santry River at the beginning of a green area in eastern Coolock (mentioned above, the link existing to reduce flooding risk). The Naniken line comtinues through Kilmore and Beaumont, taking in a flow from north of Beaumont Hospital around Ardlea Road. The river continues underground through Artane, where it caused some notable flooding in the 1950's. It flows under the part of the Malahide Road also known as Mornington Grove, near the Artane Roundabout, and passes to and through western Harmonstown. The Naniken then goes on to the cityside end of Raheny by way of a syphon under the railway cutting. Sometimes

Ballyhoy River (or **Stream**) hereabouts – though the eponymous road, of modern development, is a little distance away – the Naniken runs through old Raheny glebe lands (under a closed-up lane beside an early modern apartment complex, Rosevale) and under Howth Road by way of the buried Ballyhoy Bridge to come into the open at the boundary of St Anne's Park. There it passes sports playing fields and Dublin's Millennium Arboretum, the public tennis courts and the par-3 golf course (see left). It next touches the city plant nursery;

in this vicinity there was once a pond and a small cascade, but no trace of these is now visible. After, the river flows along a valley reshaped on the orders of the Guinness family, under several rustic-style bridges, one with a small passage, one with a track above, a passage below and a "hermit's chamber", and three simple crossings. A metal bridge over the valley – to (see right) a "temple folly", which in turn leads to a mock Herculanian house (*domus*) – was both set-up and taken down in modern times. The composite house has been restored in recent years, including its tiled flooring. In earlier times, the river was also bridged by a whale's jawbone, later auctioned off.

The river continues coastwards, passing through the remnant of a shallow pond / marshy area, overlooked by a grotto folly, and then passes through a simple sluice. It feeds the park's only extant pond, the Duck Pond, by an offtake channel, while the main line continues alongside to the west. The pond, which holds islands to shelter birds, is overlooked by a watchtower (which once sat on the roof of the estate's demolished "big house" before being relocated here) on a bluff - and features a replica Pompeian Temple of Isis (see below left).

The Naniken forms the boundary between the civil parishes of Raheny and Clontarf for the latter part of its way through the modern St Anne's Park, and at the coast. The small river finally comes to the western part of the "lagoon" between North Bull Island and the park (see below right), and the beginning of the Dollymount part of Clontarf, about a ½ km cityward from the Santry River's mouth – and at lower tides cuts a visible channel towards the estuary of the River Tolka.

The Naniken's tendency to flood, fairly regularly, at various points along its course, most notably in the Kilmore area, at Maryfield Road and around Rosemount Avenue, led to progressive culverting, and now only the St Anne's Park stretch runs freely. However, in a bit of important news for those caring for waters, it was revealed by Dublin City Council officials in 2022 that plans to "daylight" a
section of the upper Naniken as part of the major Oscar Traynor Road housing development between Santry and Coolock were likely to proceed. Daylighting is a growing trend in watercourse management in several countries, and while previous plans to open up part of the historic Bradogue River on the new Dublin Institute of Technology (DIT) / Technological University Dublin (TUD) campus in Grangegorman were long-fingered or perhaps quietly scrapped, the Naniken plan is officially backed.

In passing the Naniken mouth, we enter the substantial Clontarf area, which has several surface water lines going to the "lagoon", and shares the Tolka Estuary, as well as holding one active but often forgotten small river. There is also a flow on the modern link road – the

Alfie Byrne Road – from Clontarf to the docklands, this near-forgotten watercourse taking in the Marino Stream and other lines, and there is more on this in the Tolka section below.

A small watercourse fed from drainage channels from north of the Red Stables in St Anne's Park used to supply a second pond (shown left) near the city edge of the park by the coast. Rivulets flowed down beside staired paths of stone slabs in a rockery area. The pond was eventually removed after encountering maintenance issues due to limited flow and passivity, but the flow, coming in from two culverts, remains, partly visible, and comes to the sea modestly near the corner of St Anne's closest to the city.

Deeper into Clontarf, there was at one time a stream running roughly parallel to Seafield Road, from the grounds of Elm View, opposite the old St John's churchyard beside Clontarf Castle. This took in one small flow from north of Seafield Road East (possibly from as far north as western Sybil Hill) shortly after Vernon Avenue, and another coming east and then south near the line of the modern Seapark Road, once passing the Hibernian Marine School (amalgamated into what is now Mount Temple Comprehensive School in another part of Clontarf). The main stream then turned south-southeast, fed Crab Lake behind Conquer

Hill, and crooked southwest and southeast to reach the sea near the wooden bridge to the Bull Wall. The lower reaches of this nameless stream were still open and mapped in the early 20th century but it is now fully culverted, its remnant flow apparently reaching the sea passing under the coast road, via a drainage line visible in satellite footage of Clontarf's coastal promenade, and outfalling via a modest flap-valved pipe (shown above right).

In the late 18th century, a small stream was noted at one of Clontarf's two historic centres of population, the fishing hamlet called "the Sheds of Clontarf", at the end of Vernon Avenue. At one time this flow was piped on to a small pier, *Weeke's Wharf*, for ships to draw from. It may have been artificial, from a reservoir - but the settlement would have needed fresh water, and this may have been a natural source. At any rate, no sign of it remains today.

Possibly connected with the above was a small watercourse from near what is now the junction of Vernon Avenue and Seafield Road (on the eastern side). This, with a tiny flow from the east, ran south a little, then went along the northern edge of western Moat Lane, across Vernon Avenue and west (now south of the sports grounds) towards Beachfield House / Tudor House (redeveloped in apartments in recent years), and flowed to the sea perhaps just west of Victoria Terrace, near the Clontarf Baths. Again, nothing of this is visible today.

- We end this section with the **Wad River** (*Abhainn an Bhaid*), forming from streams (some now former) in western Poppintree in the broad Ballymun area, then losing water to the Claremont Stream (see the section on the Tolka system below) before taking in a twin flow from Balbutcher and Balcurris. After a stream comes in, south from Stormanstown, the main flow volume is taken in a long culvert, known as the *Wad River Diversion*, from central Ballymun south to the Tolka River at the western end of Griffith Park.

The Wad's historic main line, joined by a flow from Coultry Park and Shangan, goes roughly southeast, parallel to Collins Avenue, under the hidden Doyle's Bridge where Beaumont Grove meets Beaumont Road, and under Collinswood. After taking in a flow from the Grace Park area (once the upper Marino Stream reaches - see late in the Tolka section below), the Wad passes under the buried Donnycarney or Scurloge's Bridge, an occasional flood site (hidden under Malahide Road near the crossroads at Donnycarney, but with parapets still visible, and a naming plaque). The region from north of Larkhill to Donnycarney was until fairly recently known as Clonturk Valley.

The river then goes southwest, crossing Collins Avenue and running through school grounds, passes under Clanmoyle Road, where it sometimes causes flooding, and enters the lands of Clontarf Golf Club. Here it goes close to the railway, takes small inflows, runs well south-east of Mount Temple School, still in culvert (there used to be ponds in this vicinity), and flows under the rails northwest of Killester (Railway) Bridge (it used to run in a more natural line at the other side of this bridge). Some work was done in the Clanmoyle Road area, and within the golf course lands, in recent times, to help reduce flood risk, though some areas remain vulnerable.

The Wad goes southeast of the former Clontarf Railway Station and under Howth Road just west of Hollybrook Park, then through the Glaslyn apartment development, and cuts roughly south across western Clontarf in culvert (this stretch was sometimes called **Hollybrook Stream**), eventually passing by Clontarf Garda Station, and under the western end of Clontarf Promenade, to come to the sea (shown left and below).

Most of the Wad watercourse's line was still visible in the 1960s but very little can be seen now. The river is important for drainage for several areas and preparations to better manage flood risks, including the opening of a second outfall in Clontarf, were progressed during 2013, following a study released in 2009; the work included expansion of the estuarine outfall.

The Rivers and Streams of Dublin (City of Dublin, Fingal, South Dublin, Dún Laoghaire-Rathdown)

River Tolka *(An Tulc(h)a, Abhainn na Tulchann* - "The Flood")

Once joining a much wider Liffey mouth but now reaching the sea separately just to the north, between Dublin Port and Clontarf, the **River Tolka** (historically *Tulechan, Toulghy, Tullaghanoge* and the *Toulchy Water*), Dublin's 2nd river by volume (3rd including tributaries, as the modern Dodder has a greater exit flow), is a slow-moving water, falling only 140m over a 33.3-kilometre main line). The Tolka has a history of slow-building and infrequent but potentially severe flooding in parts (reaching over 200 metres wide in western Co. Dublin on occasion, for example). Salmon returned to the river in the early 2010s, after long absence, and there are also wild and placed brown trout, wild sea trout, otters and other fauna.

The Tolka rises discreetly in Meath, west of Culmullin Crossroads, northwest of Pelletstown and Batterstown, with an early branch incoming from Merrywell. The upper course, tending south / southeast, lies close to the line of the old Dublin - Navan railway and the R154 road. With small rural tributaries, and bigger near Black Bull, the Tolka passes just east of the new railway station at Pace, and runs roughly parallel to the N3. Below Loughsallagh Bridge and a little downstream of Dunboyne, it is joined by the multi-branched Castle Stream. The next part of its course passes through land with many streams and field drains – flooding is a recurrent risk from Dunboyne, Clonee and Littlepace through Damastown and Mulhuddart.

The Castle Stream's many branches originate in such townlands as Staffordstown, Cushinstown, Culcommon and Beggstown, and then near Dunboyne it meets several small channels and runs east just north of Dunboyne Castle, then southeast under Rusk Bridge. With flows incoming from each side of the old railway line, it comes to the Tolka.

The Tolka flows towards western Clonee (left) and then along the edge of the village towards the M3, largely out of sight. Historically there was a southern channel just east of Clonee, and then a millrace from the eastern edge of the village, with the river proper winding north and east. The Tolka, here about 5 metres wide, has been partly re-channelled in modern times, to flow along the first part of the former millrace, then turning northeast, crossing under and then running alongside the Dublin-Navan road. A stream joins from the north in this stretch, coming from Portan and beyond, and linked to several field drains. Shortly after, the Tolka, running north of the N3, passes into County Dublin and the main **Clonee Stream** branch merges in from culvert, northeast of Clonee itself, by the N3 slip road. This flow comes from southwest of the village, and approaching past Summerseat, splits into branches curving around Clonee. The stream's main branch, running along the county boundary for a short stretch east of Clonee, and marking the end of developed lands, used to flow into the tailrace east of the village cornmill, but this has been infilled. The stream now flows from fields into a culvert, passes between the 2nd and 3rd houses on the eastern approach to the village and then goes under the N3, joining the Tolka directly from a ½ km culvert. Whether any of the other Clonee Stream branches still flow, never mind reach the river, is unclear. The Tolka continues northeast then southeast through Damastown, an area which has seen considerable industrial development.

Streams and field drains join from the north in Damastown, and to the south a stream from Littlepace - the latter largely culverted following massive housing development, but with its final stretch still visible. Next tributaries are two left bank waters, the Pinkeen Streams, sometimes on old maps one or both shown as a "Pinkeen River."

The (**Western**) **Pinkeen Stream** flows south and east from by a well at Growtown (near Black Bush) and from Porterstown in Meath, and meets a stream from the townland of Killester, goes by Patrick's Hill in Folistown, and Causetown, then in Stokestown goes south and east. It runs roughly south from Ballintry, and on through Calliaghwee and Powerstown and southeast to eastern Damastown, where it joins the Tolka in an area of factories, adjacent to the historical Clonee millrace's end. It has many linked field drains.

The **(Eastern) Pinkeen Stream** (shown right) runs from Priesttown (its sources lie near the line of the early Ward River) and Ballintry to Tyrrellstown, picking up many small inflows. Having passed Macetown House and flowed between Base Enterprise Centre and Plato Business Park, it merges into the Tolka running southeast along a treed line some way upstream of a small cut-off bridge just upstream of Mulhuddart Bridge.

By the small bridge mentioned, a stream comes from Castaheany via one of two areas called Hansfield in the part of Co. Dublin adjacent to Clonee and Huntstown, in culvert but with its line mostly marked by green space above. It rises by Manorfields Drive, and crosses a "green corridor" and skirts north / northwest by Castlegrange. Cutting north-east, it turns east opposite 17-19 Hansfield, and flows on east with the green space to Gleanealy and Castlewood, then under Huntstown Avenue, re-entering a "green corridor" and flowing northeast to the M3 by Ashfield Park. It then runs through Blakestown, down to Mulhuddart village, where there was a mill in the past, and merges into the Tolka at a culvert outfall (see right).

The Tolka continues in a largely-canalised channel through Mulhuddart (shown left and right below), and on through the northern stretches of Greater Blanchardstown.

Warrenstown Stream, running south past Blanchardstown Institute of Technology and Corduff, is the next tributary; it has had its lower channel reshaped in the last decade.

The stream rises between Tyrrelstown and Ballycoolin, by Blanchardstown Corporate Park (opposite the former site of Cruiserath House's gate lodge), and winds south to the roundabout north of Warrenstown House (formerly Courtduff), where it takes a tributary. This latter comes from branches between Coolen and Grange (within what is now Ballycoolin Business Park), beyond Snugborough, which meet near the junction of Ballycoolin Road and Snugborough Road and the combined flow runs west.

The Warrenstown flows southwest from Warrenstown House, past Buzzardstown, and to the west of Warrenstown Green and Park (this part has been severely straightened). The stream then passes through a fringe of trees to reach the Tolka.

The Tolka then passes under Blanchardstown Road North.

A further stream, now culverted until near the Tolka, joins just downriver. It appears to rise under Corduff Grove, and runs under the community centre and playground beside Blackcourt Road (which it parallels); there was once a pond starting from the play area site. Turning south it runs through a green space, passing a well site, and emerges into the open in a wooded area in the Tolka valley. It follows its historic line towards what was a twin-channelled part of the Tolka; today the Tolka has only the southwestern line, while the stream has taken over the remnant of the northeastern line.

A curving channel around the former Coolmine House grounds (now featuring schools) used to continue to the northeast alongside where the Westend Village development now stands (as with the Castaheany-sourced flow above, the line of this channel is reflected in that of adjacent green space) and wind on northeast to reach the Tolka just upstream of Corduff Bridge; it is unclear how much of this stream is still "live" (such watercourses, even if their main water source(s) become lost, sometimes continue as drainage ditches).

The Tolka then runs south-southeast for a little - and loops under the Navan Road to run "behind" the main street of Blanchardstown's core village area. As it winds through a hidden and rather wild green area it is joined by a tributary from two southern branches. It then bends back towards the north, and returns to an easterly line north of the village, where there was once a millpond and mill. It flows along the southern edge of the grounds of Connolly Hospital (formerly the James Connolly Memorial Hospital).

A stream comes south from Abbotstown, with western and eastern branches, and passes the Swan Pond (I could find only memories of a second, smaller, waterbody, Fiddle Pond). It crosses into hospital grounds in a "managed" channel, turns sharply to the east, then south again, running down to the Tolka.

The Tolka then flows roughly southeast, and travels in canalised form under the M50 orbital motorway (see left below), near where the Royal Canal runs east-west in a raised concrete channel. It continues through the very rural Dunsink area (right below), near the former Royal

Observatory, now Dunsink Observatory and in the care of Ireland's Office of Public Works and the Dublin Institute for Advanced Studies. In Scribblestown it is wide and slow-moving (see right).

Scribblestown (or ***Cappoge***) **Stream**, flowing mostly in the open, rises north of Abbotstown, near the site of Cappoge Castle, and, having crossed under the M50, passes part of Dunsink, going directly east, with some ponding, to the south of Cappagh (National Orthopedic) Hospital, north of the capped Dunsink landfill, then bending south and meeting a small tributary which has come east and north. It turns south, flowing through Scribblestown Park in a sharp valley, and other Scribbletown lands eventually on the flat to meet the Tolka opposite Pelletstown, within the remodelled Tolka Valley Park.

Also in this part is a narrow canal, a momento of the formerly substantial millrace running to Cardiff's Bridge (much shorter than the original millrace, which was largely infilled in the course of modern park works). Two other small streams, of unclear source, flow north into the Tolka within the park, one from beyond River Road and the modern Rathborne development, one from the meeting of a buried southern line and a flow from a marshy area on the park boundary. The river itself flows on slowly to Cardiff's Bridge (from the name Kerdiff, not the city; see below), and the site of a former ironworks.

Another tributary used to join the Tolka on the right below the bridge, coming from Great Cabragh via Little Cabragh, but this is no longer clearly visible, although there is an outfall from a culvert into the river in the vicinity.

The Tolka continues in a linear park (see below), which can also hold floodwaters if needed.

Within the park area, **Finglaswood** (or **Mount Olivat**) **Stream**, almost entirely piped, comes from southwest Finglas, near the schools south of Cappagh Road, and flows via housing areas (where its course has been altered) to merge via a wetland in the park (the City Council made an "Integrated Constructed Wetland" here some years ago (see left), based around a small pond which had tended to pollution in recent decades), just upstream of Finglaswood Bridge and a golf facility. There were once ponds at Mount Olivat, and Finglas-wood House ("King James's Castle") stood by Savages Lane.

The Tolka continues fairly directly east, meeting one of its bigger tributaries some distance south of Finglas village.

The **Finglas River** begins as **Kildonan Stream**, itself forming from multi-branched flows from Grange and Kildonan. Flowing south and east, to Mellowes Park, it takes in a stream from south of Meakstown (sometimes **St Margaret's Road Stream** - branches lie either side of St Margaret's Road) and runs on partly in the open. Coming to the centre of Finglas (to which it probably gave the name, which means "clear stream"), it is

culverted, going under Church Street and Wellmount Road; the course was further straightened when the Finglas Road became a dual carriageway. Crossing to the eastern side of the road, it again runs in the open, taking in two small flows, one of which historically gathered water from
at least four source lines in the Johnstown area of Finglas, west of Ballygall. The Finglas then goes back into culvert before crossing Old Finglas Road. Just downstream, a weir guards access to an overflow line, and the two lines join the Tolka just west of Finglas Road (shown above).

 Shortly after receiving the Finglas, the Tolka runs east under the road at Tolka Bridge, going on north of Prospect (Glasnevin) Cemetery, towards the National Botanic Gardens. The river, with an old millrace splitting off to the south and then rejoining, is a feature of the Gardens, which lie partly on its flood plain (left: the Tolka flowing calmly past a northern central area of the garden complex; above right: a flood gauge on the river near the Gardens).

(Below are more photos from the verges of the Botanic Gardens – (left to right): the canal (which is actually a former millrace) which splits off, and runs within the Gardens; the former millrace reaching a sluice gate; the tail-race after the sluice, as the water rushes back to the Tolka (downstream of the Botanic's Rose Garden.)

The river then passes Glasnevin Bridge and a weir near Glasnevin's "pyramid" Roman Catholic church (pictured at top of next page).

Just downstream of Glasnevin Bridge, what is now called the **Claremont Stream** (probably the historical *Glas Naion* or *Nevin*, from which the area may be named) joins from the north; it has a northern branch, which takes much of the water from the upper Wad River course (see pp. 28-9), and one from the west, latterly extended to Jamestown Industrial Estate. With a small water flow from the St Pappins housing area, then another, from Wadelai, it runs between the former Inland Fisheries Trust building and St Mobhi's Church, then traverses the grounds of Glasnevin's Bon Secours

Hospital underground, and goes on to reach the Tolka by way of a diversion culvert (emerging through the largely-submerged arch on the bottom right of the photo above).
On the opposite bank a multi-branched watercourse flows in. Nameless on old maps but sometimes the **Cemetery Drain**, this comes from near Liffey Junction on the western railway line, north of the Royal Canal, and flows to Finglas Road, then along it - a branch either side, the northern inside the walls of Prospect (Glasnevin) Cemetery itself. This stream, with a drainage flow from the "St Paul's" cemetery extension by the railway, turns north at the graveyard Lookout Tower and runs to the Tolka, near the "pyramid church" as mentioned, just west of St Mobhi Road (not visible but on left of the preceding photo). Just downstream, the *Wad River Diversion*, built to handle run-off from the Ballymun Flats

development and other constructions, flows from the north at the west end of Griffith Park, via a hidden stilling pond; it is visible from Dean Swift Bridge (St Mobhi Bridge) (photo looking upstream seen right, with the diversion's outfall on the right).
The Tolka flows on through Griffith Park (see below), passing a former public bathing place

(weir and steps still extant, visible above in the left photo, and in focus on the right).

The next tributary is the almost-fully hidden **Hampstead Stream**, with two branches, one from Dublin City University (near a modern extension to the Albert College building), running south through Albert College Park and the grounds of Elmhurst, and another from north of Hillside Farm (for long the last farm within city bounds), which meet at and cross Griffith Avenue, go under Home Park Road and come to the Tolka by Griffith Park. On the approach to Drumcondra there was a short broad millrace to the north, near St Catherine's Holy Well; both well and millrace are now lost (the well site may lie within the northern part of the park). A stream from behind the former St Patricks's Teacher Training College ("St Pat's", now part of Dublin City University's Faculty of Education) joins just below Drumcondra Bridge, its flow today boosted by drainage lines from some way north along Swords Road, and the Tolka continues southeast. It goes roughly parallel with Richmond Road, and passes Richmond Park football grounds, and runs on between the sites of a former corn mill and a former distillery at the end of Distillery Road, the location still marked by a weir and a footbridge. The river becomes tidal as it comes to Ballybough and the Luke Kelly Bridge (view from bridge below left) and then Annesley Bridge (below right).

The Tolka is joined by one last flow from the north, with others coming together to the estuary as described at the end of this section.

The lower reaches of the Tolka's course are in effect canalised, from Clonliffe and Ballybough to Fairview and North Strand, then between Fairview Park and East Wall, finally making a left turn to run northeast and pass under the railway (shown left below), then over

the Dublin Port Tunnel (from the port to the M1 motorway and the M50 beyond, going under several watercourses) and under the bridge carrying the main access road for East Point Business Park (shown right above). The river then comes out into its modern estuary (shown right), between Clontarf and Dublin Port. The estuary merges into the open sea beyond North Bull Island.

Note: The city reaches of the river are sometimes referred to as *Drumcondra River* in historic sources.

Note on the late tributaries of the Tolka: It's no longer fully clear exactly what elements of the final flows into the Tolka are natural, in source or course, though there were at least two streams in the vicinity. The first tidal tributary, that might be called *Grace Park Watercourse*, was, Sweeney notes, *not* a natural stream in origin but a fairly modern drainage structure, largely made to assist development of early suburban housing (it may, however, have made use of natural contours). It comes from branches from Swords Road and east, some in the southern Grace Park area, and joins the Tolka ¼ kilometre above Ballybough Bridge, downstream of a tiny former millrace.

Three other waters share a diversion to the Tolka's estuary (sometimes informally called "Clontarf Bay"). The first once went along the line of Phillipsburg Avenue into the estuary at Philipsburg Strand but after land reclamation (work on what became Fairview Park began before 1910 and ended circa 1930) and walling of the Tolka's channel, was taken into a collection pipe, running northeast. The next flow, which Sweeney calls the "*Middle Arch watercourse*," began with a stream from north of lower Griffith Avenue, expanded to provide drainage for the original Marino housing scheme. These two watercourses were together diverted to the middle arch of the rail bridge over the Tolka. The collection pipe today goes on along the line of Fairview Park's boundary for its latter two thirds, taking in the residual flow of the Marino Stream.

Marino Stream was, and in piped form still is, a natural watercourse which flowed from the Hampstead area of upper Glasnevin via High Park, and filled ponds in the grounds of Lord Charlemont's Marino House. It reached the sea just a little east of the railway - its line is still visible in property boundaries behind the first few houses on Clontarf Road - when the estuary came further west, prior to the reclamation which formed the land on which Fairview Park, City Council sports facilities and the Alfie Byrne Road now exist.

The uppermost reaches of the stream have been diverted to the Wad River (described earlier), the remnant now running from south of Grace Park housing through the former Marino House lands, and, with diversions and surface drainage additions, to Marino xCrescent. One more small inflow is taken from the surface water drainage of the lower Howth Road, then the combined waters travel under the DART line and municipal sports facilities (where the Dublin City Traffic School used to be), passing under the Alfie Byrne Road on the Clontarf side of the Comhaltas Ceoltóirí Éireann (CCE) cultural centre *Clasac*. The piped stream comes to the sea at the foot of the embankment, in the inner estuary, about $^1/_6$ km from the Wad River mouth (the outfall lies within the area of Tolka Estuary shown below left, and is only visible at certain tidal levels, as shown below right).

The River Liffey *(An Life, An Ruirtheach)* *(**Anna Livia**)*

The **River Liffey**, Dublin's chief watercourse, has many tributaries, some with broad systems of their own, as it turns from mountain to bay over more than 125 km (some sources refer to lengths of up to 138 km). A true mountain stream at first, falling over 300 metres in its first 12 kilometres, it slows greatly in its middle parts. We will meet many Liffey tributaries, though we may not catch all (there are 90+ named flows *before* entry to the traditional County Dublin – approaching 60 in Co. Wicklow, more than 30 in Co. Kildare). Some are intermittent bog streams, some substantial permanent flows with rocky beds, and some of their naming is uncertain or confused, notably is long less-populated mountain areas.

The river rises in the Liffey Head Bog, a blanket bog in the Dublin/Wicklow Mountains - in the saddle between Kippure and twin-peaked Tonduff - and this locale is often mapped as "the Liffey Source." Specifically, a "bog pool" is often shown as *the* source, but the river really forms less precisely, in sodden peat over a wide area, gathering from multiple small flows. The bog holds many flushes (seepage areas), and is also source for the Dargle and Avonmore Rivers, Luggala Stream, and streams flowing to the Glencree River, while the Dodder's source is in somewhat different terrain, "around the side" of Kippure.

The Liffey headwaters gather and run northwest from the western side of Tonduff, and a unified stream crosses the Military Road between Glencree and the Sally Gap at the curious construction of multiple pipes known as Liffey Head Bridge (see left).

Note: EastWest mapping suggests that the stream to Liffey Head Bridge may be known as <u>Uisce an Fhéir</u>.

The very young Liffey begins to gather tributaries as its course runs southwest to form the third side of a triangle with the Military Road and the road from Blessington and Kilbride to the Sally Gap and beyond (R759).

The first tributary, from south of the restricted road to Kippure peak, may be *Tromán Ata*, the next possibly *Miley's Brook*. Passing the Hollow of the Tents on the western slopes of Kippure, a strong stream may be *Tromán Mór*; this part of the early Liffey may sometimes be *Cruckan Brook*. Further along, several right-hand side intermittent flows come south from Eskernaclough-more and one from Askanatriglagh, and one left-hand inflow cuts west at Shranamuck. After the river crosses the R759 (see right), it

meets a tributary formed from two flows, one from the direction of Sally Gap (Barneballaghsilurnan) and one from the southern part of the Liffey Head Bog at Knocknafoalla, crossing the Military Road just before the Gap. While smaller than the nascent Liffey, this is a serious addition, and on some historical maps, its source was sometimes mapped as a "Liffey Head". The stream from Knocknafoalla is the <u>Meadow Brook</u>. Having crossed the Military Road from the Liffey Head Bog (left), it gathers multiple bog streams; given water volumes and length, Meadow Brook may be the best name for the whole tributary too. The other branch, from sources at Ballinastoe and just north of the crossroads at the Sally Gap, is perhaps the *Asnabarney*.

Note: The name <u>Quarry Brook</u> is sometimes seen in maps of this area and may refer to the Asnabarney, the later Meadow Brook, *or* alternatively the whole stretch of Liffey from the Military Road to the R759.

The Liffey takes in what may be *Grace's Brook* from the north and flows on east to the Coronation Plantation, where it finds Lugnalee Brook on its left (tributaries: the Askakeagh, Carriglaur Brook, Carrigvore Brook) and just after that the Sraghoe (or *Sraughoe* or *Glossnavillogue*, which is also the name of a small tributary) (two views shown below) from north along the Kippure Forest boundary.

A streamlet comes from Kipppure Forest to the north, the Glasnaslingan from Glenflugh Flat and Furry Hollow to the south, and the Cransillagh from the north. Athdown Brook (or *Adown Brook*; see left) tumbles south next, down from Seefingan, with a tributary from Seefinn, followed by the Shaking Bog Brook coming north from Brockey Bog on Asknahick. Further on in the Glen of Athdown two streams meet and shortly after run in at Ballynabrocky above Ballysmuttan: Ballylow Brook - with tributaries including Lugduff and Lavarnia Brooks - and Ballydonnell Brook - sources include Boleyhemashboy (or *Ballyhemusboy*) Brook (and in turn Glenvadda Brook, with tributary Whitebog Brook, and Glenagoppul Brook with its tributary, Parkbawn Brook), Dealbog Brook (with tributaries Tent Brook and Tramhongar) and Luggcullen Brook.

Right after, the Liffey turns sharply to run northwest as the first of three streams from Ballynatona, the eponymous Ballinatone Brook, joins, followed by two more small flows.

Scurlock's Brook joins at Ballysmuttan Bridge (the memorial Ciaran Jones Bridge, see right) on a minor road west of the main road. Then, below Cloghleagh, in Church Glen, the Shankill River (pictured below) joins - it forms from Seechon Brook and The Slade (which has, in turn, the Cloghoge Brook tributary).

The Liffey soon after turns west, and then southwest, and in the Glen of Kilbride, three streamlets join.
The next substantive inflow is the **Brittas River**, which flows partly in Co. Dublin. Part of its course within Wicklow is also called the **Kilbride River** and part the **Lisheens River**.
The Brittas joins the Liffey just below

Ballyward Bridge and its story is more complex than those of most Liffey tributaries. It runs north from the northern slopes of Seechon (Seahan) Mountain, taking flows from below Ballymorefinn and Black Hills, then turns west, coming down by Ballinascorney Gap. At Pennybog it meets a stream from the southwestern face of Knockannavea. As a many-branched stream joins from Ballinascorney, the Brittas enters Wicklow, and then, just after a small tributary joins from west of Butter Mountain (see right), its flow was for many years divided at a sluice at Aughfarrel (the sluice no longer works, at least not in normal flow
conditions). The divided watercourse flows west in near-parallel channels, with the man-made northern line re-entering Dublin and turning north just before Brittas, where it was routed to provide strong input to the upper Camac River approaching Brittas Ponds (this channel once brought most of the early Camac volume but seems to carry little water most of the time today). The Brittas course proper turns south, taking in a stream coming south from Slademore and Glenaraneen, one from Cupidstown Hill and a multi-branched flow from Lisheens, between Butter Mountain (this has no dairy-related connotations, but is rather from *Sliabh an Bhóthair*, Road Mountain) and Dow(e)ry Hill, before it runs past The Lamb, and flows south by (Manor) Kilbride village to the Liffey.

Ballyward Brook, from Sorrel Hill via Ballynasculloge, with a tributary joining just before Stoneyford Bridge (formerly the Ston(e)y Ford), and passing the edge of Ballyward Forest, merges in from the south on a meander, then a streamlet from eastern Threecastles.

We then reach Poulaphouca Reservoir (or Pollaphuca, etc., or "Blessington Lakes"), formed in the valleys of the Liffey and the Kings River, and their merged flow (most of the lake area is actually in the Kings River valley). Before the structure was formed (probably recreating an ancient lake), the Liffey ran southwest to Blessington, then south, merging with the Kings River - at that point, in the past, probably the larger flow - north of Baltiboys, with the combined course following the Kings River line northwest, then winding south and south-west to Poulaphouca Falls, and continuing towards Ballymore Eustace. The artificial lake has three main divisions: northeastern and southwestern ones, and a larger part to the southeast. The county boundary of Kildare meets that of Wicklow halfway down the southwestern section, south of Russborough House and its maze. The Liffey enters the reservoir at Threecastles, and exits near Ballymore Eustace.

The "surviving" tributaries (following the formation of the large artificial lake) first met are Woodend Brook on the eastern shore, from Lug na gCon or Blackamoor Hill, and Sorrel Hill, and Black Brook on the northern shore, from Tinode. Many remnants of longer flows, such as Dwyer's Brook from Rathnabo, what may be Srothansoillaghe from Sroughan, and Tromawn Brook and others at Kilbeg Glen, enter on the eastern side. Ballynastockan Brook / Cock Brook, formerly joining the Kings River and now falling directly into the reservoir, is followed by streams that once flowed into itself, such as Fraughan Brook, Oghill Brook (collecting a tributary, Troman Brook), Ballyknockan Brook and Annacarney Brook (with Quainty Brook), and a stream from Lockstown to Blackditches. Short flows (no names traced as yet) also join on the northern and western sides of the northeastern section, e.g. at Crosscoolharbour and Haylands, and on western slopes falling to the southeastern section, as at Boystown.

The Kings River (or Owenree) itself enters Poulaphouca at its southern end. It rises in the Wicklow Gap and collects Annalecka Brook, then what is sometimes the Ballinagee River, formed from high flows joined by Glasnagollum Brook (and Gowlan Brook) from the east and then <u>Glasnadade Brook</u> from the west. After bog streams from the south, a larger water joins via Garryknock Bridge, then <u>Glenreemore Brook</u> (with <u>Asbawn Brook</u>) and <u>Knockalt Brook</u>. Further north, the <u>Douglas River</u> (with two tributaries each called <u>Trether Brook</u>) and then the <u>Little Douglas Stream</u>, flow in.

After the Liffey leaves Poulaphouca, it winds into County Kildare, west, north and west again. Streams enter the Golden Falls run (expanded due to the hydroelectric plant dam), from Whitebog via Killerk, and from Silverhill. A millrace used to be taken off just upstream of what was then the Golden Waterfall, and supplied the large Woollen Mill in Ballymore Eustace; only a remnant of its lower stretch remains. In Ballymore Eustace, below the former mill, the Liffey passes under Ballymore (or Liffey) Bridge, in the area once called The Strand; there has been a bridge here since at least 1599. The river then receives a spring-fed streamlet from the south and at the western end of the town, the <u>Seasons Stream</u> flows in, having passed under Pinkeen Bridge on the Naas Road.

Later Kildare tributaries include <u>Lemonstown Stream</u> at Ardenode (major source the <u>Toor Brook</u>, with additional flows from Coldstreams and Wards of Tober), the <u>Brook of Donode</u> at Co(u)ghlanstown and a small nameless stream at Gaganstown (historically Yagoe). After meeting streamlets east of Logstown, the Liffey reaches the Kilcullen area.

Upstream of Kilcullen Bridge (today usually simply Kilcullen, since the people of the original hill town, which became known as Old Kilcullen, largely moved Liffey-side after a bridge was made centuries ago), <u>Kilcullen Stream</u> comes in from the south (the lower reaches of this watercourse are known as *Mill Stream*). The main line of this stream flows via Yellowbog, from sources south of Cartersbog, including Gilbinstown east of Usk. A millrace was historically taken off to drive a mill at Kilgowan. Near Milemill a small flow joined from the west, then a millrace began to the east of the main stream. A stream winding west from the Gilltown area joined the millrace, and the two lines ran north in parallel, taking in small spring-fed flows. What was the second of three channels linking the millrace and the main stream now marks the end of the former millrace, southwest of the road and Newabbey House. Previously the mill channel went east of New Abbey Cemetery and having passed an eastern overflow channel to the Liffey, powered New Abbey cornmill, the tailrace running north to the river (little trace of mill, tailrace or overflow channel remains). The Kilcullen or Mill Stream itself flows to the west of the ruins of New Abbey and meanders to reach the Liffey about half a kilometre below the former mill outfall.

The tributary from Gilltown originates from several branches, including one from Kennycourt, and feeds Gilltown Lake. It once drove a sawmill by Gilltown House, and supplied a short southern millrace and mill pond for a corn mill in Milemill.

The next tributary is a <u>Pinkeen Stream</u>, flowing by the southern boundary of Castlemartin Estate to the Liffey, having risen from springs east of today's Cnoc na Greine Woods, and then having crossed the community pitch and putt course.

From central Kilcullen the Liffey flows roughly northwest, with bends. Two small flows join either side of the M7, downstream of Castlemartin, coming from either side of the old long avenue of Castlemartin Demesne; one of these used to feed from Crabtree Well.

At Athgarvan is a small millrace, and another survives at the site of Great Connell Abbey. Just beyond, a stream joins at Kilbelin (no name has been traced but it has been the subject of improvement works in recent years), then a bypassed bend in the slow-moving Liffey is visible south of the town of Newbridge.

In northern Newbridge, <u>Sexes Stream</u>, ultimately from Morristownbiller, joins. Historically rising halfway from Blackberry Lane to Morristown Biller Road, it now seems to begin somewhere near the railway line and Lakeside Park, supplies the pond between Lakeside and Dara Park, crosses and runs near the railway, and emerges from culvert northwest of Newbridge Station. It then crosses back under the railway and having received a smaller stream from Piercetown, curves north towards Sexes Road, and reaches the Liffey south-west of St Eustace's Church. Two fairly modern channels run part-way between this stream and Mooney's Stream, described below, and there may be a connection.

Just above the weir by Roseberry or Rosberry Commons, a former millrace takes off water on the Liffey's western bank. Shortly before it returns to the river to the north, this channel intercepts <u>Mooney's Stream</u>, which used to run direct to the river. The stream, from western Roseberry, used to be fed by a series of springs and small streams, and once started near Sarsfield House, but has become foreshortened over the years. It runs parallel to Mooney's Road, and under the railway before winding north and east.

The Liffey then winds west and east as it continues north; a ditch on the right marks where a small stream used to flow from Oldconnell. Another stream flows in from the west in northern Rosberry, having passed under the railway; the most remote line of this comes from Hawkfield, and it is connected to an extensive field drainage network. Downstream of Newbridge several streamlets enter, including flows from by the Naas and Corbally Grand Canal branch, and Herbertstown (or Corbally) Harbour, which lies at its remotest point. Streamlets join at Tankardsgarden, from the west, and Coolreagh, from the east. Next is what may be called the <u>Pinkeen Stream</u>, at Morristown, coming north. This flow, with roots by Corbally Harbour, meets a former bend of the Liffey, now cut off, and runs counter-clockwise with this to the river proper, opposite Barrettstown House. In the same area, a streamlet joins from the west at Barrettstown. A little downstream a bigger stream flows in, having just taken in a tiny flow from St Patrick's Well in the riverside churchyard.

At Thomastown, two streams run south to the Liffey, east and west of Morristown Lattin across the river. A little to the east, in Yeomanstown, a short former millrace heads east at the weir at Skeagh Bridge, while a stream joins from the south, and the Liffey passes Victoria Bridge before the tailrace rejoins. The old cornmill by Victoria Bridge, perhaps in origin the oldest mill extant in Kildare, has been partly restored in recent years.

There are traces of a small flow at Moortown then, as the Liffey bends sharply north at Newhall, and just upstream of a group of islands, a tailrace joins from a millrace taken off a stream at Cormick's (or Comick's) Bridge to the south; the stream itself flows in in western Halverstown, as the Liffey meanders further. This stream, running partly parallel to the N7 west of Naas, has multiple source lines, including drainage cuttings.

<u>Awillyinish Stream</u> (or *Annislingh Stream*) forms from multiple branches, the two biggest coming from Donore to the north and Lattensbog to the west, and joins the Liffey south of Caragh (Carragh), just upstream of Caragh Bridge (possibly the narrowest Liffey bridge). The river then turns sharply to the east and takes in small channels from Castlekeely to the north and Halverstown to the south. After further meandering the Liffey, having passed a former offtake channel running from Coy Ford, comes to Osberstown. Here it receives a tributary from the south, with lines from near the twin sites of Castle Rag and Jigginstown House (or Castle) beyond Ploopluck Bridge in Naas's northern suburbs.

At Waterstown a stream joins from the west, and the Liffey is crossed by the Grand Canal in the impressive Leinster Aqueduct. Several more small streams flow in, including one from Sallins, and then more near Millicent (southwest of Millicent House, just below Millicent Bridge, and an eastern flow opposite Granny's Wood).

At Clane, the <u>Butter Stream</u> (*Butterstream*) joins. This comes from branches west of Loughanure, and runs roughly east to Butterstream Commons, passes Halfmile Bridge and continues east, partly culverted, picking up flow from several springs. It eventually turns to flow south east near the northern end of Millicent Road, passing a *bullaun* or *cup stone* and running just west of the ruins of Clane Abbey, where it takes in a small left- hand tributary. The Butterstream then goes south east and east to meet the Liffey just downstream of Alexandra Bridge, Clane's river crossing.

The Liffey runs northeast from Clane, then meanders to flow north, receiving a southern contribution, from Blackhall townland, on the bend. After continuing northeast and north, the river turns sharply at Wogan's Hole, to go southeast; it meets the <u>Gollymochy River</u> on the point of the bend.

The Gollymochy comes from northwest and takes in a flow from the north, from the Pale Ditch and from west of Clongowes Wood College, with one source line from Mainham, near

Queen Buan's Grave, and others from Betaghstown. After a tributary from a local source, another line from the Clongowes Wood area flows in, and then streams, from Capdoo and further south, and, coming south, from branches and springs in Irishtown.
The Liffey receives further Irishtown flows, a mix of stream lines and field drains, and meanders on east, north and northeast.
Downstream of Irishtown House, on the boundary of the Straffan Demesne / Kildare Club ("K Club") lands, a straightened stream flows into the Liffey. This stream has multiple sources, notably several branches in Ovidstown, Baybush and towards Barberstown, but also a number of springs closer to the river.
After receiving another left bank tributary, the Liffey passes around Inishmore island and approaches the broad Straffan area (the village proper lies a little north of the river) and a weir (which used to help power a turbine) and Straffan Bridge. Immediately below Straffan Bridge a stream joins in the treed area on the right (a branch of this stream used to come in above the weir also), flowing from springs in Turnings (near Shortwood) and Littlerath to the south. It still follows its historic line, except where it crosses the southern "K Club" golf course, where ponds now draw from it. This marks the beginning of a stretch of the Liffey commonly known as "the Jungle" due to its wild, over-shadowed and overgrown nature.
Below Straffan, within "the Jungle," a major tributary, the _Morell River_, flows in on the right. The Morell system is one of the largest watercourse structures in the Liffey watershed, and notably several members of the system cross under the M7 in succession.
The Morell proper begins with branches going northwest from the slopes of Slieveroe, joined by a flow from Bride's Well and other Walshestown sources. It passes Beggar's End and Seven Springs, turning north in the Tipper area, to head for Johnstown. What may be a drainage line runs nearly parallel south of Johnstown, and receives _Annagall Stream_, the merged flow entering the Morell just south of Johnstown's main street, all continuing under the N7. A little north of Johnston, the _Tobenavoher River_ (the main volume of which comes from the _Hartwell River_) joins from the southeast, and the Morell goes on north, to run between Kerdiffstown House and Palmerstown House. Having passed through Palmerstown lands, the river reaches the Grand Canal, under which it runs just above the 15th lock. Two "feeder streams" run from the Morell to the canal, one from near Johnstown, to Sallins, and the Morell Feeder from Palmerstown to just below the 14th lock, in Kileenbeg. The river then passes under one Morell Bridge, and shortly after under the railway. It takes in two flows from the west, and then, by Turnings House, the accumulated _Painestown River_ from the southeast, before passing under (Old) Morell Bridge. A millrace, the line still discernible although largely dry, used to separate at the bridge, to run roughly parallel to the west, taking in some small streams and heading for Straffan Mill. The river itself passes one final Morell Bridge, after which it used to receive the tailrace from Straffan Mill (only the faintest traces of this remain). It then bends to flow into the Liffey at an angle, perhaps a ½ kilometre below Straffan Bridge, depositing sand to help make a shallow stretch. We now trace the key Morell tributaries.
The Annagall Stream rises in southern Rathmore, near Rathmore School, and passes Daffy Lodge. It meets a southern tributary in Baysland townland, passes the site of Newtown corn mill, and is divided into two channels, the northern of which "dead-ends" in Furness. The southern line continues past Forenaughts House, with some flow possibly diverted to a southern secondary line to the Morell. The main flow heads northwest towards the confluence near Johnstown mentioned above.
The Hartwell River rises in several branches between Rathbane and Punchestown Upper, some in wooded area, and flows northwest and west to reach the site of Segrave's Castle, wind north of Rathmore Moat and turn northwest. At Broguestown it receives a tributary rising in multiple springs in Rathmore. In this vicinity the main river sometimes seems to be called the _Rathmore Stream_. The Hartwell continues northwest to Burntfurze, where more spring-fed lines join. Passing southwest of the Hartwell Castle site (and a former millrace), it turns west, meets a small flow from Toberton and becomes

the Tobenavoher River, passing under the bridge of the same name and winding towards the Morell. It forms a lake in its final stretch, which has been much modified (it once formed a key feature of Palmerstown House's demesne, and fed a millrace, but is now reworked within a golf course).

The Painestown River is the last major Morell tributary, most of the volume of which comes from the Slane and Kill Rivers, so we will consider those two flows also.

The Kill River rises southeast of the pictureqsue village of Kill, passes the village's motte-and-bailey, goes under the main street and through a small park, then north, northwest and northeast. It then meets and joins the Painestown River. The small Painestown has come from the southeast, from Kilwarden and Porterstown. It meets a tributary and passes under a back road behind a service station on the M7 on the Dublin approach to Kill, and then goes piped under the M7.

The combined flow goes on roughly north to pass under Painestown Bridge, just west of Painestown House, and soon after meets and takes in the Slane.

The Slane River comes from branches in eastern Kildare, merging at Oldmill. The more northern branch winds west from Johnstown in the hilly area west of Saggart while the southerly branch rises in hillside springs in Cromwellstown and Cupidstown and heads northwest. From Oldmill the Slane runs northwest, passing under the N7 and later flowing under Tuck Mill Bridge and then beginning to curve west. Still in Tuckmilltown, the small river receives a tributary from Oughterard to the northeast, in the area where the grave of Arthur Guinness lies in a small graveyard at the foot of a ruined round tower. Soon after, the Slane turns south and runs west of south, passing Finger Post Bridge, and then White Bridge near Bishopscourt House before making a sharp, almost ninety degree, turn at Mill Bridge. It then passes Old Mill Bridge and takes in a small stream before running north of Painestown House to meet the Painestown / Kill River flow.

The unified Painestown River runs on north, receives a stream in Alasty and curves west, crossing the railway at Three Aqueducts, going northwest, and curving round to flow into the Morell a little southeast of Turnings House.

The Liffey turns sharply north, bends east, picks up a stream from the north, and then goes on broadly northeast. On a sharp meander at Friarstown a stream comes in from the south, with its main source passing the sites of a monastery, moated house and castle, as well as Melon's Bridge, all in Whitechurch, before taking in a small flow from near Straffan Station, and another from south of the railway line. Another tributary, from streams and field drains, joins a few metres downstream, and the Liffey runs on north, east and north again.

On the approach to Celbridge, Pausdeen Stream joins at Newtown, then the Liffey passes the weir which managed the millrace of Temple Mills (the millrace has survived the mill's passing), and undulates northwest through Celbridge. In the town centre, a second weir, sometimes called Vanessa, used to control the flow for Celbridge Mills. Modest streams from south and north join, the most significant of these being the small Toni River, from Griffinrath, which runs south, east to the north of Pickering Forest, roughly east to Oakleypark, and on into the town. It used to reach the Liffey in two parallel branches, one of which remains visible, the outfall a little to the southeast of Kildrought House. There is also an area of rapids.

Just downstream of Celbridge, Crodaun (*Croudaun*) Stream joins to the left, having run just west of Castletown House, where it was diverted by the famous Conolly family. The Crodaun comes from several branches, one from the Sally Well in Moortown, and several being channels in Crodaun Forest. It historically came out of culvert below the works yard (see left), to fill a long narrow pond by the front lawn, and, after this, two channels. One of these runs directly to the Liffey, passing under two "rustic" bridges, and one runs to the east (in a "ha-ha" or hidden ditch), passing north of the temple folly, to meet a small flow from woodland to the east (almost reaching the Barnhall Stream, mentioned below), and then flow to the river west of the estate's ruined icehouse and east of a little bathing house at the Liffey's edge (this channel also runs under two bridges). The pond was "absent" for many years, but has been restored (see below), with the dam

which sustained it being preserved and reinstated, and the pond-and-channel system returned to operation in 2013. The temple folly (Mrs Siddons Temple), ice house, well and bathing house were also to be worked on.

The Donaghcumper Stream (or *River*) joins from the right, at Donaghcumper Demesne, having passed under Rock Bridge and ornamental estate bridges, and with a key source at Dangan. This small stream supports some trout.

Shinkeen Stream flows in a little further downstream, via Hazelhatch and the site of St Wolstan's Abbey and its tower. Just downstream is a hidden weir, St Wolstan's, sometimes called the "ghost weir" as it is only intermittently visible; this used to divert water for Newbridge Corn Mill, to the south of Newbridge town.

The Liffey then meets the modest Barnhall Stream. This stream, whose history is unclear, flows by both Barnhall Rugby Club and the industrial facilities of Hewlett Packard (HP), going roughly southeast to the river.

The Liffey passes under New Bridge and widens into Leixlip Reservoir (sometimes Lake), which has a wide section, and then narrows, passes under the M4 motorway and goes on to Leixlip Dam and the ESB generating station (see right).

A little along the northern bank of the reservoir, Ryan's Stream runs in, having travelled southeast from branches on HP lands at Parsonstown, and having gone under the R404. Its farthest source is at Kilmacredock and one line passes near the famous Wonderful Barn.

The Rivers and Streams of Dublin (City of Dublin, Fingal, South Dublin, Dún Laoghaire-Rathdown)

Just after, another stream flows in, coming southwest and having passed the M4.

A stream joining in two branches on the south bank, at Coneyburrow and Backweston Park, comes all the way from beside the Grand Canal at Balscot, having run under the railway at Stamcumny Cottage. The western branch passes Backweston Farm and the eastern passes under the runways of Weston Airport. In a section between Backweston Park and Leixlip Demense, the impounded Liffey's right-hand side enters County Dublin.

At the western end of Leixlip, passing the Leixlip Castle lands, the busy Rye Water (or Ryewater River, or Rye River), from Meath via Kildare, merges in. Beginning with branches from west and south of the churchyard at Agher, and from Baconstown and Ardrums, and joined by a line from Summerhill Demesne, the Rye runs southeast, and from a point between Ferrans and Killeighter parallels the Royal Canal (the river course was modified, at least in this vicinity, during canal construction), passing close at Ferns (Ferrans) Lock. Below Balfeaghan Bridge, a tributary joins from Newtownrathganley, and the river runs southeast, then south to Kilcock, remains close to the canal for a little and then heads north of east, picking up small tributaries and eventually skirting Maynooth's centre to north and northeast, and receiving the Lyreen.

The Lyreen River forms from branches from Corcoranstown, Grange and Ballygagan, (it once powered a cornmill in its upper reaches, which are sometimes partly labelled Loughtown River) and approaches Maynooth from the east. It takes in Baltracey River, from around Donadea, including a flow from the lake and walled stream in the forest park there. The Baltracey in turn meets a modest water from south of Loughtown, then the Clonshanbo River, then a tributary from Moortown and, at Baltracey House, another stream from Rathcoffey and Raheen.

The Lyreen passes under the Royal Canal a little southeast of the canal's Jackson's Bridge, and shortly after, a millrace separates, east of Laraghbryan churchyard, to run east a little north of the river proper, right alongside the R148, on the approach to Maynooth. Shortly after the millrace begins, it intercepts a small tributary coming south from Crew Hill. Both millrace and river flow within the South Campus (the older part) of St Patrick's College / NUI Maynooth. When the new university library was being built, serious consideration was given to having the Lyreen running inside as a feature. The plan was not pursued, and the river now runs just to the south, and the millrace just to the north (crossed by two bridges), of the library. The millrace merges back into the Lyreen by the Manor Mills (Shopping) Centre, built on the site of the mill it once powered, and the river curves and runs northeast towards Mill Street / Dublin Road.

The next tributary to join, just before the Lyreen passes under Mill Street, is the Joan Slade or Owenslade River (shown right, this name refers especially to the stages north of the Royal Canal), or Meadowbrook River. This normally rather modest flow has a disproportionate ability to cause flooding, even on the motorway. The river, also sometimes Rowanstown River in middle reaches, forms from branches from near Bryanstown, which

merge north of the curiously-named Cocked Hat Wood and Maguire's Wood, the line on-going via Donaghstown and western Taghadoe, and Rowanstown. Just before passing the M4, the river takes a tributary from Windgates via Taghadoe and Dowdstown. It then passes under Brookfield Avenue and parts of the Meadowbrook development. Taking two small flows from the west, the Meadowbrook passes the canal near Bond Bridge, and flows along Parsons Street to the west, then goes under the street and swings around to run north. It goes under the bridge leading from central Maynooth to the main college entrance, and travels east of Maynooth Castle to meet the Lyreen.

The Lyreen continues northeast and another stream runs in from the south. This normally small water comes from branches from Moneycooly (south of Greenfield) and Ballygoran (northeast of Griffin Rath Manor), the merged flow passing through the Rockfield development, where a small western tributary joins, then crossing the Grand Canal beside Mullen Bridge and Maynooth's railway station. It continues underground parallel to Straffan Road, comes into the open at Manor Court and then goes into an old culvert known locally as "The Tunnel" and passes the market / court house square and Main Street to join the Lyreen as a rather discreet piped flow.

The Lyreen turns away from Maynooth's Main Street, runs northeast, meets a flow from Crew Hill and Timard branches, and goes on to join the Rye at Mariavilla, just upstream and northwest of Kildare Bridge; there are ponds in the angle of the two rivers.

The Rye then runs east, to the south of Carton House, and takes in the modest but grandly-named Glashnoonareen River (historically also *Offaly River*), which marks the

county boundary for a stretch, just upstream of Carton Bridge. The kilometre of Rye Water ending at this point used to be much wider, and today there are two former islands in the vicinity; a stream from a spring in the Railpark area of Maynooth also joins. The river then widens into The Lake (of Carton Demesne), which angles north-west to south east, and features a decorative boathouse and several islands (see left). Some way to the north is another lake, "The Sheet of Water," probably artificial, whose outfall comes to the Rye. The Rye itself flows over a weir at the end of the main lake, by Carton's Shell Cottage, meanders past the site of Bride's Well and passes around Ham Island opposite Cobbler's Cave. After several small islands, one of which is crossed by Black Bridge, the Rye flows roughly east, then turns south east, and takes in a tributary cascading from Shaughlin's Glen. The latter stream, with sources in a wet area north of Bogganstown, also takes water from Shaughlin's Well.

The Rye continues a little to the northeast of the massive Intel facility, then turns south and east. It finds tributaries from north and south, and crosses the Royal Canal and the railway. Flowing east, southeast and roughly south, it passes through northern Leixlip, flows under Rye Bridge (pictured right) and, just north of Leixlip Castle,

turns southeast (see left below) and then turns sharply east to run parallel to the town's

Main Street, and join the Liffey as the latter itself turns east. Near the turn, a pretty cascade joins from Leixlip Castle's demesne (in background left above, close-up right above), alongside another piped flow.

At the downstream end of Leixlip, Silleachain Stream joins, flowing from north of Confey and on between the ruins of Confey Church and the site of Confey Castle, and crossing the railway. Running to the west of the substantial St Catherine's Park, passing Leixlip Fire Station (stream and station shown right), and going a little east near the end (there was formerly a Liffey millrace here), it joins the Liffey discreetly.

Almost immediately after, the Liffey crosses fully into County Dublin.

The Liffey flows roughly east across County Dublin, with a meandering course, including a couple of sharp bends, especially in the western parts. Along the way, it takes in 14 named tributaries (some with substantial systems of their own) and at least 10 unnamed:

- **Tobermaclugg Stream**, from branches between Milltown and Clutterland south of the Grand Canal. A tributary joins from the west by Airlie Heights, north of the Tobermaclugg Holy Well site, coming from Backstown and Stacumny and passing Backweston Park and Weston Airport. The Tobermaclugg flows north within the Adamstown development area, largely beyond the parts developed so far, and passes, in the angle between Millstream and Old Cornmill Roads, just east of the Spa Hotel, to reach the Liffey just west of Lucan.

- **Griffeen River** (sometimes *Griffin River*) (*An Grifín*), passing through Lucan's village centre and flowing into the Liffey after crossing a strip of old demesne lands, long home to the residence of the Italian ambassador. Rising on Saggart Hill at about 400m altitude, it flows north and tributaries join from the hills south of Saggart and Rathcoole, and from the Athgoe area. Within parkland in Esker South, west of Tullyhall, **Kilmahuddrick Stream**,

which has come west, north and northwest, flows in (alterations were made to its course in recent decades in the course of work on the railway). Later two streams from southern Lucan also join. Much of the lower course of the river (see far left) is surrounded by linear park, running roughly north past Griffeen Avenue via Esker Road and the Lucan By-Pass, and turning west and narrowing towards Lock Road. A small stream joins (near left) a little downstream of the crossing of Lock Road, coming from Esker via a lake on private land south of Mount Gandon. Within the linear park, the Griffeen is crossed by King John's Bridge, reputedly the oldest extant bridge in Ireland (the old bridge over the Boyne at Trim may be of like antiquity). The river then crosses the centre of Lucan in a rock-floored channel (below left) with an inscribed bridge at one end (below right).

Parts of the river were historically known as *Racreena River* and, near the area of the same name south of Lucan, *Esker River*. A historical branch course splitting off just north of the Grand Canal and flowing via Finnstown and past Somerton House still apparently reaches the Liffey somewhat upstream of the main Griffeen course, near the ruined oratory in former Lucan Demesne lands between Dodsboro and Lucan.

- A small left bank stream from two branches in Laraghcon, passing Clanaboy House

- A small stream from the western end of Luttrellstown Demesne, south of Luttrellstown Castle; it is not clear if this is natural or fed by a human-made offtake from the next mentioned water. It passes through the former (Shackleton) Anna Liffey Mills complex.
- **Westmanstown Stream**, coming roughly east from Meath via Pass-if-you-can, turning and crossing Tinkers Hill and entering Luttrellstown Demesne. After meeting a streamlet from the north, it forms a long lake in the middle of the estate, then passes south into a wooded glen and receives what is sometimes called **Kellystown Stream**. The latter, entering the demesne near the golf clubhouse, features three small lakes and crosses the golf course. The stream flows roughly southeast in its Luttrellstown glen, then cascades, and runs south under Lower Road to the Liffey.

A small watercourse comes to the north side of the Liffey at the Wren's Nest. Beside is a weir, well-known in canoeing circles, and opposite this a three-kilometre-long millrace begins. This parallels the river to the south, past Waterstown to western Palmerstown, where it drove a mill; it was historically crossed by five stone bridges, with a right-of-way running alongside.

- Two small streams flowing north, intercepted by the just-mentioned millrace, one from Larkfield via Cursis Stream (which may be named for it) and Quarry Vale, and one in Palmerston Lower, and a third small flow coming east through Palmerston.

- **Diswellstown Stream**, running south to the Liffey past Knockmaroon Demesne, near the site of the once-famous Diswellstown Ragwell (covered, with a marked stone near). The stream comes from branches from Carpenterstown and the Castleknock village area. It flows substantially in the open (see left) but out of easy sight in an often-deep valley.

- **Glenaulin Stream** (sometimes **Pound Lane Stream**), from west of Palmerstown, passing through Palmerstown's Glenaulin Park and Ballyfermot's California Hills Park, and taking a diversion from Blackditch Stream (see below) and a small flow from western Ballyfermot. The main modern outfall is hard to see, at Glenaulin Weir. Though normally modest, the stream has on occasion flooded the road to Lucan badly enough to make it impassable.

On the Northside, we now begin to meet the Phoenix Park watercourses.

- **Furry Glen** (or **Knockmaroon**) **Stream**, now coming from beyond Farmleigh, where it feeds a pond east of the house (below left). It then runs south of White's Gate into the Phoenix Park, and on through the grounds of Ordnance Survey Ireland. Where north-south and east-west valleys meet, a stream joins from the west, coming from north of Mount Sackville (it used to flow from a pond, no longer extant)

and passes Baker's Well (above right), from which it takes a modest flow; it is sometimes known as the **Baker's Well Stream** or **St Brigid's Stream**. The main Glen Stream passes through a marshy area and forms Furry Glen Pond (pictured at top of next page) before running out of the park and down to the Liffey. Oddly, early mapping shows no stream to or from Farmleigh, which initially had no pond either - on such maps, the Furry Glen Stream appears to have begun at a "cistern" high in the glen. It is not clear if the other parts of the stream were artificially formed in the 19th century, or simply previously unmapped. The little Baker's Well Stream *was* historically mapped.

There is also a streamlet at the eastern end of the Glen, visible just south of the high road, and later passing under Martin's Row and on to the Liffey.

At Chapelizod there is a millrace north of the river proper. Next is **St Laurence Stream**, from Ballyfermot, going parallel to the line of Lynch's Lane to the south side area by the river, once St Laurence Village, now

marked by St Laurence Road, crossed in culvert. This stretch of the river (see above) features several boating clubs, and the War Memorial Gardens approaching Islandbridge.

- Two streamlets running from the hilly southern park area, one visible at Knockmary just west of the Cheshire Home boundary and then running southwest, and one nearer St Mary's Hospital (these are linked in a drainage channel part-encircling the hospital lands).

- Several small culverted flows coming north from parts of eastern Ballyfermot.

- The **Creosote Stream**, from west of and going through Inchicore (Rail) Works in branches. Pollutants from the works gave the modern name (an unusual one but comparable to the Kerosene Stream in Moscow, for example, or Kerosene Creek in New Zealand). Having passed Sarsfield and Con Colbert Roads in culvert, and taken in a small tributary, the Creosote comes into the open towards the western end of the Lutyens-designed Garden of Remembrance, and to a deep part of the Liffey via "the coffin" (see above left). A tributary which used to run west via the Nash Street area is now diverted, via Tyrconnel Road, to the Camac (discussed later on).

The Islandbridge millrace, beginning at the ancient great weir north of the war memorial gardens (which also marks the normal tidal limit of the Liffey), runs south of the Liffey for a short stretch, and can be heard, and distantly seen, rejoining the river just west of the bridge of the same name (previously Sarah or Sarah's Bridge; view from it shown right, towards the viaduct carrying the railway line to/from the Phoenix Park Tunnel), before Usher's Quay.

The Rivers and Streams of Dublin (City of Dublin, Fingal, South Dublin, Dún Laoghaire-Rathdown)

- A stream gathering several small flows from the southern reaches of the Park.
- **Magazine Stream**, the southerly of the two main Phoenix Park streams, rises beyond Castleknock Gate and runs into the park, south of, and partly parallel to, the central road. It passes Quarry Lake (below left), which was made from three smaller ponds.

It is not clear if the stream flows *through* the lake today but 19th century mapping appears to show it flowing in on the west side and out to the south. There is a winding channel leaving Quarry Lake southwards, broadening into a swampy area (above right) surrounded by soggy grassland, from which a rough stream joins a ditch to the south. The stream goes on to traverse the edge of Oldtown Wood(s), crossing Tinkler's Path (below left). With small inflows, it goes around the lands of the USA's ambassador's residence (the former Chief Secretary's for Ireland's Lodge), partly in "ha-ha" ditches, and reforming from two channels and tending southeast, passes the Papal Cross area. From there on it goes east and south, taking in a small stream from near the cricket grounds, and going down in the open, near the Khyber Road (sometimes this part, flowing first to the west, then to the east, of the road, is called the **Khyber Stream**) (below right). It then flows down Whitebridge Hill, passes below Magazine Fort, and comes to the Liffey in the upper tidal reaches at Islandbridge, opposite a former mill site.

- **River Camac** (*An Chamóg*) (historically **Camoke**, **Cammoke** or **Kamoke**, **Cammock** or **Cammack**, some upper reaches sometimes **Swift River** or **Brook**, parts near Drimnagh and Crumlin **Crumocke** or the **Crooked River** and possibly sometimes the **Slade** or **Slade More River**). Although apparently unknown to residents in its upper area in the 19th century, the Camac name *is* referenced in documents from at least the 1300s and 1500s.

Dublin city's 4th river by volume, the modern Camac is partly the result of 19th century modifications. Existing flows in the Slade Valley, the highest of which still runs from Mount Seskin (c. 370 m), and near, were boosted by a major diversion from Liffey tributary the Brittas River (discussed earlier); the diversion took in the old highest flow. The main volume of the river now runs from the the Brittas Ponds (see right), at and below Glenaraneen, which occupy what was originally the site of a south-flowing stream and boggy land ("Bog Larkin") on low ground.  The northern pond was made first, to provide a head of water for mill operations, and with the building of a further embankment (almost 10m high), and a dam with a sluice channel, the southerly pond was formed later, as a deep fish pond (one source says that both ponds were used for milling *and* fishing). A channel from the "enhanced" Camac fed the upper, southern, pond, to which a stream once also flowed in the glen called *Slád na bPlumpóg* or *Stawdnabrumpoge*, since impacted by a road alongside, and now normally dry. At the northern end of the first pond a stream - possibly named for its little valley - still appears to flow in the glen called the **Ferny Glinn** or **Glanarawny**. A channel then runs north through the partly dried-out but still marshy site of the northern pond, joined by the *White Stone Slade*, now mostly dry, from the west, and the "pond channel" rejoins the main river line a little further north. There is also a little glen called *Slad na Raoife* or *Slád na Riaibhche* somewhere in the vicinity but its location is no longer clear. Further tributaries include the two distinct streams known as the **Two Slades** (*An Da Shlád*) coming down from the west and a streamlet, which may have been known as **Ath Collop** (*Áth Colpa*), joining between these two, from the east. Then there is tiny **Toberach** also from the east, by the hilly road to the Crooksling medical facility - the last section of which was the **Stream of the Neighing of Horses**.

Bending west on the western slopes of Verschoyle's Hill, the Camac receives a substantial flow from Saggart Hill, and later streams from Coolmine and Rathcoole.

A millrace, which may have been known as "Tobarach," heads away towards the eastern side of Saggart, and rejoins, from beyond Slade Castle. A small stream, and then a drainage line, join from the west, and the river passes under the N7 and heads northwest, with three streams joining from the south in succession. The Camac runs alongside Clonlara Road, close to the edge of Baldonnell Aerodrome lands, and the first of the three streams flows in. This is **Boherboy Stream**, coming from the valley east of the Slade of Saggart, where it is the **Corbally Slade River**. The second stream, larger, is the **Narahan Water** or **Brownsbarn Stream**, that comes north between Corbally and Lugmore, past Fortunestown, and the third, the **Killeen Water** (sometimes today **Fettercairn Stream**), approaches from around Cheeverstown (its line historically began where Glenanne Park lies today, just south of the source of the Poddle, running west and northwest towards Cheeverstown Castle, then west and northwest again). The first two traverse parts of the broad "Citywest" development, while the last lies a little to the east. The Boherboy joins the Camac a little north of the

Citywest interchange on the N7, and the Narahan west of the Kingswood interchange, while the Killeen actually runs through the latter interchange and flows into the Camac as it goes east, expanding into Kilmateed Pond, from which it exits a little west of north. The Camac goes north and then turns east, passing through the Corkagh area, where it features in the large public Corkagh Park, formed from an old estate, with traces of old mill works. The park also contains a small arboretum and has provision for flood interception. The river proper runs northeast, with water taken off to feed fishponds (one shown above right) and stream lines, which largely reunite west of St John's Green. Also, a small stream rises within the park, at what was Sruleen Well, and flows north to the Camac. Approaching Clondalkin (this stretch was sometimes **Clondalkin River**), there is a large millrace that splits off southeast of Cherrywood Avenue. The Camac is culverted in the town centre, until north of Orchard Road, before running northeast to the Nangor Road, which it parallels to the Fox & Geese and Bluebell areas (there used to be notable meanders in this area, but they have been straightened out in modern times.)

The next major tributary is **Robinhood Stream** (historically the **Coolfan River**), which has some uncuIverted stretches (including at Newlands Cross Cemetery and within Robinhood Industrial Estate), and comes broadly northeast from Newlands Demesne (one source a well there), passing Ballymount, where there are ponds, historically receiving multiple tributaries from Ballymount, Walkinstown and Greenhills (part of one still visible), and joining the Camac from the south in a culvert under John F Kennedy Avenue and Drive (it used to have two confluence points north east of Drimnagh Lodge).

From the north runs the **Gallblack Stream**, formed in the joining of the **Blackditch Stream** (from Rowlagh via Wheatfield - where there is a partial diversion to Glenaulin Stream, discussed in the Liffey section - and Killeen, with parts of the course diverted to facilitate housing and other developments) with the **Gallanstown Stream** (from Neilstown, seen passing in branches under the Fonthill Road and near the canal water filter beds). After passage under the Grand Canal, the last stages of the Gallblack are diverted in a wide (3.5m each way) culvert, reaching the Camac underground near Bluebell Cemetery.

The Camac runs in a further culvert either side of the Naas Road, then into the Lansdowne Valley, through which it runs (see both photos to left), and the original name of which, meaning "Crooked Glen", gave Crumlin its name.

Drimnagh Castle (or *Bluebell*) **Stream**, from towards Greenhills (still supplying the moat of the castle, pictured below), is next to merge in, by a former millrace, in the area of the Lansdowne Gate development. This is in turn close to the last-joining major original tributary, the **Walkinstown Stream**, also from Greenhills and Walkinstown but slightly further south. The Walkinstown, which occupies a northwest-oriented branch of the Lansdowne Valley, also takes flows from the western end of O'Brien Road and the Crumlin Crossroads area.

The Camac passes under the Grand Canal, 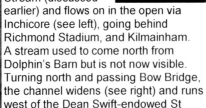 takes a diversion from the Creosote Stream (discussed earlier) and flows on in the open via Inchicore (see left), going behind Richmond Stadium, and Kilmainham. A stream used to come north from Dolphin's Barn but is not now visible. Turning north and passing Bow Bridge, the channel widens (see right) and runs west of the Dean Swift-endowed St  Patrick's Hospital, and Dr Steevens' Hospital too. Behind Swift's Hospital was the offtake for an ancient channel, known as the *Camac* or *Usher's Island Millrace*. This important, almost certainly artificial, line appears to have left the river just below Bow Bridge, heading east; it is

further discussed in the Poddle section below (from p. 58). The Camac finally flows into a large culvert (over 6.5 m wide, holding sewer pipes too) and under Heuston Station, coming to the Liffey (see left) at a broad arched opening nearly 100 m upstream of Seán

Heuston Bridge (the former King's Bridge), in line with platform beginnings. The arch can be seen at lower tides.

The river flows mostly in the open, and having been Dublin's most polluted major watercourse, is now clean in most reaches. Not so much as one of the many millraces its system historically contained survives.

- **Finisk Stream** (or ***Viceregal Stream*** or ***Zoo Stream***), from eastern Castleknock, north of Castleknock Gate, and going southeast roughly parallel to the Phoenix Park's central road, Chesterfield Avenue. It passes the Machine Pond by Mountjoy Cross (no clear regular flow between the stream and this former quarry is known). A southerly flow from the former Ashtown Trotting Track (now the Chesterfield housing development) may join underground from the north in this stretch (the White Fields area). The Finisk then flows to the west of the official residence of Ireland's President, Áras an Uachtaráin, near Civil Defence buildings, through the area known as "The Wilderness." It passes north of the Áras, previously the Viceregal Lodge. In the 19th century, the main channel was still visible here, along with a partly parallel channel to the north, beyond which lay the Poor Man's Well. The Finisk then expands into the Áras Pond, a modest remnant of the former presidential pond - historically the Fish Pond of the Viceregal Lodge; the bulk of pond and surrounds were transferred to Dublin Zoo (one of the oldest Zoological Gardens in the world), where they now form the acclaimed *African Plains* area. Moving from presidential to zoo lands, the second, main,

division of the old Fish Pond is, at an area of 2.8 hectares, the largest lake in Dublin city. The Finisk flows on to the older part of the zoo, past the former site of the Phoenix Park spa well. There was previously a single zoo pond, which is now divided into three sections (lower pond on left). At the western end of the lower pond, the Finisk passes a sluice into a culvert and flows to the east near the zoo's main entrance.

A nameless underground overflow, which may have formed naturally, or been made to relieve ground flows of water, enters the Phoenix Park Rail Tunnel just before the People's Gardens and reaches the Liffey just downstream of the rail bridge.

The Finisk runs on under The Hollow beyond the historic Tea Rooms, supplies the People's

Gardens' Pond (right) by Bishop's Wood, and then the smaller

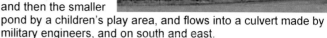

pond by a children's play area, and flows into a culvert made by military engineers, and on south and east.

Earlier, there were two substantial upper ponds, with at least two exit channels from the dam at the eastern end of the lower, and three channels (two with ponds), all fully merged in the vicinity of the then Soldiers Hospital, after which a pond on the slopes of the hilly area at the eastern end of the park sent one more stream to flow in.

The Finisk crosses Infirmary Road, tending towards Montpelier Hill, and passes a former foundry, later a fabric shop. It joins the Liffey less than 20 m before Seán Heuston (formerly King's) Bridge (see the bottom right of the photo to left), just upstream of the line of Steevens Lane on the southside.

The Liffey flows on through the city centre (see photos below), past embankments erected over the centuries (the past shape of the river was quite different, and it was a good deal wider in places). Many bridges span it, worthy (and the subject) of whole books.

It is often asked why there are not more stream and drain outlets visible in the Liffey quay walls. In fact there are dozens of openings, but many are only visible at lower river levels and few see regular flows to the Liffey. The explanation is simple - Dublin was an early beneficiary of intercepting sewers, laid inside the quays by 1906 and capturing most (54 out of 57, per one report) lines which once drained to the river. One interceptor runs from around Parkgate Street to Eden Quay and then crosses under the Liffey to meet, at Burgh Quay, with the other, from around Steevens' Lane and Heuston Station; the combined flow goes on to Ringsend. While long stretches of quay are pristine, the working outfalls either carry notable flows such as the Poddle (and maybe the Bradogue), or act as overflows when the intercepting sewers are heavily loaded (on the combined sewer overflow [CSO] model). Later interceptor sewers in Dublin include the Grand Canal Tunnel Sewer, also taking input from Lucan and Clondalkin, and from greater Blanchardstown. Other drainage features include the tying of Tolka catchment drainage and North Strand lines to the Eden Quay interceptor, the Liffey Tunnel from East Road to Ringsend, the Dodder Valley Sewer, and the sub-sea connections of northern (from in front of St Fintan's High School on the Sutton – Baldoyle coast) and southern (from Dún Laoghaire) drainage networks to Ringsend - and piping from pumps at the site of some springs into the sewers (but Dublin's issues in this respect are modest, while some cities, notably those with underground rail systems, such as New York, would have widespread serious problems within hours, without constant high-volume pumping).

On some maps of medieval Dublin a tributary appears to meet the Liffey in the Oxmanstown / broad Smithfield area. There is no stream known here today but, for example, an "Oxmanstown Stream" was conjectured, based on topographical factors, in the "Rivers of Dublin," running southwest from western Cabra and passing Blackhorse Avenue, Oxmanstown Road, Stonybatter and Queen Street, to reach the Liffey somewhere near Mellows Bridge (Queen Maeve Bridge, Queens Bridge). There is, as it happens, an outfall just downstream of that bridge, which may or may not relate.

- **Bradoge** or **Bradogue River** (also known as the **Bradok**, **Glascoynock** or **Pole Water**, or **St Michan's Streams** historically) comes from Cabra and was the only significant tributary north of the original Dublin, providing water to the Oxmanstown settlement.

The upper reaches are not traceable today but the primary source was a little southwest of the junction of Nephin Road (once "Blind Lane") and Ratoath Road, with the river running east. A branch comes south from the junction of Ventry Road and Broombridge Road, additionally fed by a spring. Two parallel lines run east south of Faussagh Road, meeting near Inver Road. By the junction of Drumcliffe and Carnlough Roads, the bulk of the upper course's flow is diverted from Cabra to an outlet to the Tolka by Finglas Bridge, and the

Bradogue line goes on east under a rough green space and the Connolly-Heuston railway line, to approach Quarry Road south of the historic site of the house known as Beggs Boro'. It then turns east and follows, roughly, the line of Offaly Road, where it used to take in a stream from the north near Bregia Road, headed southeast across Leix Road to the Cabra Road, and met a channel coming northeast from Annamoe Road. The Bradogue line continues along the property line between the back gardens of Charleville Road and Annamoe Drive, meeting a flow coming northeast across Annamoe Park and coming to the North Circular Road between the Charleville and Royal Terrace segments. The river's course is reflected in the curving rear line of Grangegorman Upper properties (behind St Elizabeth's). The line continues through Dublin Bus's Broadstone base (previously it lay behind an engine shed and turntable within the Broadstone Railway Terminus) and a very old diversion channel takes much of the flow off to the south within the eastern part of what was remade as the Grangegorman Campus of the Dublin Institute of Technology (now the core campus of Technological University Dublin). This was for decades Grangegorman Hospital, and before that the broad areas was home to an asylum, a workhouse and at least three hospitals.

The historic main line of the little river tends southeast and east, and near where it crosses Constitution Hill used to pass under a harbour on a Royal Canal branch line, and the Foster Aqueduct. Somewhere in this vicinity, it reputedly was bridged by a large stone, hence the name "Broadstone" for the area. The Bradogue then runs alongs the northern edge of the Kings Inns lands, formerly part of the garden of the Abbot of St Mary's Abbey, and parallels Dominick Street to join Henrietta Lane, following the lane southeast and then southwest, along the western side of Bolton Street and via Green Street and Halston (possibly originally Half Stone) Street, between Little Britain Street and Cuckoo Lane, and across Chancery Street, which was once Pill Lane (there was historically a multi-channelled mouth area called "The Pill" or "The Pole" by a wider Liffey, with a watermill belonging to St Mary's - but this muddy area was long ago reclaimed, after the walling-in of the Liffey.) The Bradogue line heads across colourful, remodelled, Ormond Square to the Liffey at Ormond Quay Upper, but it seems that residual flow is generally captured by either or both of a sewer at

Ormond Sqaure and the intercepting sewer within the quay walls (which goes on to near the East Link Bridge). Still, there is a valved outfall by Arran Street East, a little upstream of Capel Street, and it is possible that a little of the Bradogue may sometimes be seen reaching its parent there (see right of centre in photo to left). (Sweeney in "The Rivers of Dublin" wonders if the river may originally have flowed on in a gentle curve to Ormond Quay Lower, but no sign of such a course is found today.) Meanwhile, the ancient diversion channel from Grangegorman passes the site of a former women's prison which later formed part of the Grangegorman medical complex, and downstream the Legion of Mary's Morning Star Hostel, after which it historically passed Richmond Hospital in two branches to meet North Brunswick Street (once called Channel Lane). The culverted river diversion passes along Red Cow Lane, crosses Smithfield on the east and comes to Arran Street West, with any flow not captured by the interepting sewer outfalling to the Liffey at Arran Quay.

The Bradogue has been thoroughly integrated

into the municipal sewers (unusually - not counting combined sewer overflow inputs - with both surface water and foul flows, at least at Grangegorman). Plans to separate and perhaps even "daylight" (unculvert) parts of it have been discussed in light of the development of the TUD campus at Grangegorman but it as of 2022, with the initial phase of campus development complete, had not proceeded.

- **River Poddle** (*An Poitéal*) (historically also the ***Pottle, Podell, Poddell, Puddell, Salagh, Soulagh, Glasholac***, etc.), a truly historic watercourse, and the one on which the Black Pool (Dubh Linn), which gave Dublin its name in English, formed. The early settlements which became Dublin formed near the confluence of the Poddle with the Liffey, where there used to be a sizeable bay, one of several inlets on the lower Liffey. Limited parts of the upper Poddle course have been modified but the line there is largely as it was, while the lower course has been altered and interconnected with other watercourses, natural and artificial, over many years, from at least the 13th century to the 20th. Channels divide and reunite, and there is no absolute consensus on the details. Mostly in the open to beyond Mount Argus until the late 60's (then culverted except for tiny stretches, as at Warrenmount in Blackpitts), today a considerable part of the Poddle system is hidden.

The Poddle begins as the ***Tymon River*** in Cookstown, in the vicinity of the former Jacob's Biscuit factory and beyond Tallaght Hospital; the exact source is not clearly visible but much of the early course can be seen. It runs by the Technological University Dublin Tallaght campus (formerly IT Tallaght) and flows to the north of Tallaght village, passing an athletics

track and Bancroft Park (it also used to take in a stream from near Tallaght Priory), then runs under Greenhills Road and on through the Tymon North area. Having gone northeast past and between schools, it turns north and northwest within Tymon Park, feeding three smaller artificial ponds, and one larger (see left), then goes northeast under the M50 motorway. Having flowed through more human-formed ponds, including Tymon Lake (below left), it exits to the east (below right), running parallel with Limekiln Road, and passes from Greenhills, under Wellington Lane, to Perrystown and Kimmage.

The Poddle was joined until recently near what is now Greentrees Road by the first of the "*City Watercourse*" channels, a major diversion from the River Dodder at Balrothery north of Firhouse, that actually supplied the majority of the downstream Poddle flow (see the Dodder section for details). This diversion was made, not later than the 1240s, to provide drinking water for the early city of Dublin (the Liffey was always salty in the city centre, the Stein River small and off to the east of the core city - and the Camac was perhaps too far west of the original south bank settlements, though it is likely that some water *was* brought from that river via what later became the Camac / Usher's Island Millrace). Only vestiges of the ancient, and for long vital, connection - largely taken over by drainage lines - now exist at either end.

There is a mandate of April 1244 in which the Justiciar of Ireland directs the city authorities to look into the question of an urban water supply. Some believe this to be the basis for the City Weir / Watercourse arrangements, but there is evidence that works were begun earlier by the Abbey of St Thomas, and modified in cooperation with the city in 1244-1245.

From Kimmage on, the river is known exclusively as the Poddle. In Kimmage a historic open channel supplying the lake at Terenure College (previously Terenure House), which drained to a tributary of the Swan River (see pp. 76-77), was replaced in the 1930s by a culverted flow, the *"Lakelands Overflow,"* to the west end of the school's lake (shown below left).

A culvert line, once accessible for passage for the not so tall - and also sometimes noted as *Lakelands Overflow* - goes on from the lake under the Rathdown development. It emerges in Bushy Park, falling in a picturesque open cascade by the east end of the lower park pond. The overflow used also to supply this pond but is now channelled onwards to the Dodder (above). The lower pond is instead today fed by a small cascade from the upper Bushy Park pond, supplied in turn by an offtake from the Dodder.

The Poddle is in the open (sample stretch shown to left) until past Kimmage Manor, where it passes through the religious complex, including going under one building, and is then culverted until near Kimmage Road West. It is next culverted from Ravensdale Park, where there was a millrace and millpond, and then parallels much of Poddle Park, before running between Blarney Park and St Martin's Park, then going into culvert again near

Crumlin's Sundrive Shopping Centre, built on the site of the former Larkfield Mill complex. Re-emerging by Mount Argus Monastery, the Poddle has been divided by a shaped stone (called "the Tongue" or "the Stone Boat") for over 750 years (see right). The left-hand fork here (at what was long known as the Tongue Field or Tonguefield), that can

take up to about one third of the flow, becomes the second *"City Watercourse"* stretch, the other branch contuining the Poddle course.
(Old) City Watercourse: The inner City Watercourse branch receives water only when the Poddle runs at a certain level, and goes northeast and then north in culvert, its line passing the Grand Canal at Crumlin Road, by which it is intercepted by the Grand Canal Tunnel Sewer. The Watercourse line comes via Dolphin's Barn to an area known as the "back of the pipes" where it briefly runs parallel to the modern main Poddle course, and formerly fed two pools and filled the substantial City Basin south of St James Street. From James Street distribution watercourses ran (as discussed further at the end of this Poddle section). Prior to the construction of the City Basin in 1721, the water supply was distributed from the Great Cistern, or City Cistern, sited near St James's Gate on Thomas Street. Main distribution was to parts of the old core city south of the Liffey, but there was also for many years a supply by pipe north across The Bridge to the small but growing northern extension of the city.
The river goes on from the stretch between Kimmage Road and Mount Argus Road, partly in

culvert. When the area just downstream of the Tongue was, in recent times, developed, bridges and a pond were added, with a powerful fountain (not now working). The river then passes in the open along the eastern boundary of Mount Jerome Cemetery, from a half-hidden reach (see left) behind the Russian Orthodox Church and community centre, and then goes back into culvert to pass the lands of Harold's Cross Hospice and beyond; it does not surface again for any significant length until reaching the Liffey (two tiny sections remained visible into the 2000s, and one still does).
From the vicinity of the entrance to Mount Jerome, the Poddle course was modified many centuries ago, with, as far as can be ascertained, no remnant of the original course extant. This curiously-laid out reach of the river, as far as New Row, was known as the **Abbey Mill Stream** *for many years, and later as the* **Earl of Meath's Watercourse**.
The original Poddle course: It appears that the modest river used to run fairly directly towards the old city and its confluence with the Liffey. This would be close to today's Harold's Cross Road, then parallel to Clanbrassil Street as far as New Row. It was diverted significantly to the west, to the line of the 12th or 13th century *Abbey Stream* cutting.
The Abbey Stream / "modern" Poddle course: This goes on from the edge of Mount Jerome via the front areas of the Harold's Cross hospice lands and Greenmount (formerly a milling complex, now a small business park), and under the Grand Canal at Parnell Place. As with the City Watercourse, it passes the Grand Canal Tunnel Sewer drainage diversion line before the canal passage, with a syphon taking the river onwards, and an overflow to the sewer for times of high flow.
The main Poddle line today passes Donore Avenue (once known as Love Lane); in this vicinity a small stretch is visible (see right). It runs on towards the site of the James's Street Basin, then away again, tracing three sides of a rough rectangle. Along the way north of the canal, some water was

"taken off" at the sluice called "Roaring Meg" to supply other channels, notably the *Tenter Water* artificial one, which "cuts across" the rectangle, and now takes much of the residual Poddle flow more directly. This channel crossed the area known as Tenter Fields, where fabric was once spread to dry. The combination of Poddle lines once formed a "Donore Island." Some sources suggest a take-off prior to but more-or-less parallel to the Tenter Water, which may have been the "Factory Water".

The Poddle makes its way on through the Liberties, passing Warrenmount, where a second short inner city stretch was still visible in the 2000s, and crossing Blackpitts. The course from Harold's Cross to Marrowbone Lane was historically almost level, and then fell by 8 metres in a short span within the Liberties, enabling the operation of mills, and washing areas for tanners and weavers. By Fumbally Lane, the combination of a small stream and the Tenter Water merge into the Poddle, followed at the junction of New Street and Kevin Street by the Commons Water, over and under which the Poddle (Abbey Stream) line passed earlier. On the other side of this junction, where the line of the Coombe meets Patrick Street, there is a short stretch east of the Coombe called Dean Street, which appears to have been previously named Cross-Poddle for the river. Around here also the river resumes its early historical course, the Abbey Stream diversion coming to an end.

> The **Commons Water**, fully culverted since 1874, and passing over and under Poddle system elements, including the City Watercourse, has its main source at the former Crumlin Commons near the village, with a probable tributary from Drimnagh. The merged flow goes along Mourne Road; in later years, the residual flow appeared to begin between Mourne, Keeper and Dolphin Roads. It then runs under the Grand Canal (where its line is intercepted by the Grand Canal Tunnel Sewer) and through the Coombe (this part is sometimes **Coombe Stream**) to the New Street / Patrick Street confluence.

The Poddle's Patrick Street reach: After Dean Street, the unified Poddle line comes to Patrick Street, along which it used to run in a stream at each side, culverted long ago (the original course, further west, made an island here, on which the early, Celtic, St Patrick's Church was built, and there was for many years a watermill where St Patrick's Park now lies). The dual line layout was present until the 1920s, when the Poddle was gathered into a single flow in a brick culvert on the eastern side of Patrick Street. This runs east, traversing a large chamber (7 metres wide and 3 high) and coming to Dublin Castle at Ship Street, entering more or less at Ship Street Gate.

Flooding: The Poddle has a history of flooding, most notably in the St Patrick's Street area, and also around Tripoli and Ardee Street. This is why St Patrick's Cathedral has no crypt - water has sometimes risen to within less than half a metre of the cathedral floor, and concerns about this led, for example, to the moving of the graves of Dean Swift and Stella. Sewers were built in the cathedral curtilage to reduce the water levels, and improvements ordered in 1825 also took in the waters of St Patrick's Well. On an historical note, Henry VII required householders living near the Poddle in the old city area to avoid adding to flood risk, and from the 1660s to the 1840s, there was a Poddle Commission, operating under Acts of the Irish Parliament, which had powers, including the right to collect a special tax, to try to manage the situation. The Commissioners and their inspectors, always under-funded, struggled both with clogging of the channels and with blocking of the flow by mill operators, and manufacturers using the water to wash goods, both of the latter claiming to be operating under leases from the Earl of Meath, through whose *Liberty* the river flowed. The powers of the Commissioners were passed to Dublin's Commissioners of Paving in 1840, and later to Dublin Corporation.

Castle reaches: Having passed Dublin Castle's Record Tower and Chapel Royal, the Poddle turns sharply to the north; it was in this vicinity that the Black Pool (*Dubh Linn*, from which the name "Dublin" comes) lay. A formal space, the Dubh Linn Garden, near today's Chester Beatty Museum, commemorates the pool. As it approaches the Palace Street gate, the Poddle divides into two streams to leave the Castle, the smaller passing that gate while the larger runs towards City Hall and then turns towards the Liffey to parallel the other. The two channels flow either side of the Olympia Theatre and the larger branch then turns east to

meet with the other at Essex Street, near the Project Arts Centre. The unified Poddle goes to a grated outfall to the Liffey at Wellington Quay (shown above left), near the Clarence Hotel and Grattan Bridge.

Parts of the river's Dublin Castle reaches have been shown on television.

Related watercourses: The various watercourses, probably all artificial, associated with the Poddle and city water supply, are complex, and having been much altered, may never be fully mapped: aside from those mentioned above, they include at least the Glib Water, the Limerick (Lord Limerick's) Watercourse, the Crockers' (Barrs) Stream / Lea Brook and Col(e)man's Brook, as well as the Camac-Usher's Island flow (see also the Camac River part of Liffey section).

We will briefly meet these related watercourses in turn below.

As mentioned earlier, what is now known as the *Camac or Usher's Island Millrace* was a flow drawn off from the Camac River a couple of hundred metres below Bow Bridge, running north a little and then east towards the old city of Dublin. It probably predated the Dodder-Poddle arrangement of the 13th century, and could have been used to bring fresh water towards the early city, in a line clear of even a spring tide Liffey. It ran across what would now be the front lawn of Steevens' Hospital, crossed Steevens' Lane, and cut across the quayside lands now occupied by Guinness. It then turned to the Liffey just west of Watling Street. However, it may also have fed the next-mentioned watercourse. It operated for centuries but the upstream part was eventually filled-in, the downstream course still operating as a drainage line from Guinness lands at least in the 19th century.

Crockers' Stream (Crockers' Barrs Stream) flowed west to east across an area between Thomas Street (by "the gate at the Barrs") and the Liffey, occupied by crockers (potters) in the 13th century and known for several hundred years after as Crockers' Barrs. It may also be the *Lea Brook* appearing in early documentation, the name perhaps relating to a holding known as *Lea Land*. The stream can be seen on Speed's 1610 map of Dublin, and local topography suggests that its course was probably not natural. Latterly it joined the lower reaches of the Glib Water but before that was made, the Crocker's Stream could have continued on to the city ditch (a partial moat running north to the Liffey and southeast towards the Poddle from a high point near today's Cornmarket) and walls, or even the next mentioned watercourse.

Colman's (or *Coleman's*) *Brook* was a short channel from west of Bridge Street Lower running east, then turning north near a lane called Skipper's Alley, no longer extant but just west of the Adam and Eve Church. It was substantial enough to be navigable by some form of boat in its lower reaches, and its purpose and origin are unclear but it could have been the final part of a longer structure pre-dating the Glib Water. The Brook was culverted around 1721, and its line still at least partly functions as a sewer.

The *Glib Water* appears to have been the name from the 17th century of one or more supply lines, also referenced as *Thomas Street Water* and *City Conduit*, running from the city's water storage by upper Thomas Street (first the City Cistern, later the City Basin), flowing near Thomas Street and ultimately supplying water to the mill at the Usher's Island area and to the city ditch.

The *Limerick Watercourse* (or *Lord Limerick's Watercourse* or *the Back Course*) was an extension of the City Watercourse, running from the City Basin to St James's Gate, and on to the Liffey, east of Watling Street. It was the subject of a confrontation with Arthur Guinness in person in 1775, and of a lease made from 1784-10759 A.D. Yes, 10759... The watercourse supply was substituted by other piped lines, and it was finally cut off in 1941. On some maps it is simply shown as a late section of the City Watercourse.

The Liffey continues, passing under the Ha'penny (former toll) bridge, and then O'Connell Bridge, before receiving its penultimate tributary, one of ancient naming.

- **Stein** (or **Steyn**, **Steyne**, **Stayne**, or **Stane**) **River**, named for a Viking marker, a pillar stone (the "Steyne" / "Stein") that used to stand near where the little river met the Liffey, in a wide area of sloblands similarly named (in those time the Liffey at high tide came up to what would now be the northern edge of the Trinity College campus, on Pearse Street, in the area once called Lazar's or Lazy Hill). This small river, along with the Poddle, Camac, Bradogue, and Liffey, helped define early Dublin. The Stein was covered rather early in Dublin's development, has been much altered over time, and is now merged with the sewers.

The Stein rises south of Adelaide Road, near the site of Charlemont Bridge, and its line runs east, north-east and then north, under the former railway from Harcourt Street, then sharply northwest across Earlsfort Terrace, along the northern wall of the National Concert Hall, then behind UCD's Newman House & University Church and north of the Iveagh Gardens, turning sharply north after crossing Harcourt Street and crossing York Street. While it passes near St Stephen's Green Park, and once formed part of two side boundaries of the original, larger, Green, the Stein does not supply the park with pond water, which instead comes from the Grand Canal (at Portobello) on a gravity line. The Stein next runs by Mercer Street, having been diverted to permit the building of Stephen's Green Shopping Centre, and then heads roughly north, by Clarendon Street, Wicklow Street and the west side of lower Grafton Street, to which point it was formerly tidal (posing a danger, even a mortal one, in basements).

The last stages of the Stein pass northeast of Trinity College - the Priory of All Hallows (or All Saints), which preceded the university at what was then a location *outside* the city, had a bridge and watermill - and College Street. The Stein used to reach the Liffey near the junction of D'Olier Street and Pearse Street but the Liffey was reshaped by embankments commissioned by Hawkins and Rogerson, and the little river's line now runs northwest and north, to Burgh Quay.

No further tributaries are known before the Dodder, but TCD subsoil mapping shows a watercourse (labelled "Unknown") from the east side of St Stephen's Green via the east end of the Trinity campus, and Sweeney conjectured one more flow, from Leeson St. Lower north west to Merrion Square, then north to City Quay (he suggested "Gallows Stream" as a name).

The Liffey enters its final stages after the now-obscure Steyne confluence, passing the (in)famous Link Loop Line Bridge viaduct (see below left) and the parallel Butt Bridge, the

Custom House (above right) and Talbot Bridge, the Calatrava-designed Sean O'Casey

bridge, and the Samuel Beckett Bridge. It then receives the combined flow of the Dodder, its largest tributary, and the Grand Canal, and passes the East-Link (Tom Clarke) Bridge, widening as it enters the Dublin Port area.

The Liffey opens fully into its estuary after the Pigeon House on the south banks, and, having met the Tolka's estuary, the combined mass of Dublin's major watercourses flows to the open sea between the arms of the Bull Walls.

(Several photos of the last stages of the river are shown below, the last looking to the East Link Bridge, beyond which the channel widens and merges into the bay…)

Many books discuss the Liffey, but beyond Sweeney, especially fine are de Courcy's encyclopedia-style *The Liffey in Dublin*, with deep coverage of the city area, and *The Book of the Liffey (from source to sea)* by Healy, Moriarty and O'Flaherty, with quality notes and photos from all reaches of the river. The majority of the tributaries were not explored much, or even listed, before Sweeney (for the city area) and later my work, for reasons unknown, but there *are* multiple articles about or touching on the Poddle, and a few likewise regarding the Camac, the Bradogue and the Steyne, for example.

The River Dodder (*An Dothra*)

Today the largest Liffey tributary but once reaching the sea at a wide separate mouth west of the old city centre, and Dublin's second watercourse by typical exit volume, the **River Dodder** has a steep gradient in its early parts, and a gentle slope for most of its lower course. The main channel length is around 26 kilometres. The river is "flashy", responding quickly to rain in the mountains, and has a record of occasional severe flooding, at times even washing away bridges. The Dodder features salmon, sea trout, brown trout, lamprey and otter, and its vicinity hosts a wide range of flora and other fauna, including many bird varieties.

The definitive guide to the main Dodder line is *Down the Dodder* by the prolific Christopher Moriarty (cf. biography page), along with the City Council *Dodder and Poddle* book; no work yet goes into great detail on the tributaries.

The Dodder begins on the north-western face of Kippure mountain, with a three-branched stream often mapped as **Tromanallison** ("***Allison's Brook***"). In the highest, western-most, branch, twin gullies from around 670 metres altitude (at least one source mentions source lines from 750 to over 760 metres) are joined by a third channel. These early lines seem dry at times but by the time they meet the second main branch at about 525 metres, are wet permanently. Just above the 500-metre level, the third main branch, from a few hundred metres upstream in Co. Wicklow, merges in (some suggest that *this* branch is properly Tromanallison and the other flow simply the Dodder). The combined stream flows rapidly downhill north and northeast, collects a small left-bank tributary (no name is known) and meets at around the 410-metre contour with the active **Mareen's** (sometimes ***Maureen's***) **Brook** (*Aill Mháirín*). This latter runs roughly north from across the Wicklow border, with an eastern tributary just inside County Dublin, and then the fast-flowing **Cataract of the Brown Rowan (Roan Tree)**, also running west, joining. These merge just before Tromanallison and the brook come together at the area of confluence known as the *Head of (the) Three Brooks* (*Bun na dTrí Tromán*) or *Meeting of the Three Brooks*.

An overview of the Dodder's high reaches is shown below:

Two more right-bank tributaries, one **Trumandoo** (*An Tromán Dubh*) and the other with no name found, join the young Dodder as it flows through the Dodder Gorge (shown right below); the gorge was carved by glacial ice, not the river. The river goes on by Mary's Cliff. The early tributaries form on slopes which are waterlogged even in full summer. One flows parallel to the gorge for a stretch, then takes in a tributary, and shortly after

turns towards the Dodder and runs sharply downhill in a cascade (left and centre above), while the next flows from across a mountain road, roughly west and directly to the Dodder.

A third right-hand tributary joins after the valley has broadened out, gathering several mountain streams and drainage flows, and soon thereafter a small left-hand stream.

The next part of the Dodder course was altered in the 19th century, to improve water supply to Rathmines and Rathgar, and to help manage flood risks, both of which remain important functions (the scheme still contributes actively to Dublin's water supply today).

Originally, the Cot Brook merged with the Slade Brook, and that flow met the Dodder, which had collected further hill streams, beyond Castlekelly Bridge, well inside the valley of Glenasmole (featured in several legends, especially of Finn and Oisin), just before Glassavullaun Stream merged in and the river wound north along the valley (small Castlekelly tributary shown right).

Today much of Glenasmole is occupied by the two Bohernabreena Reservoirs. The slope of the river down to the reservoirs is 1:15 and notably the Dodder's main flow does not normally pass through both lakes, only the lower one. The upper is used to supply drinking water, and as the water of the upper Dodder is often peat-stained, the engineers decided to divert the main river flow around it, and to supply the reservoir from the gathered flow of smaller streams.

As a result of the engineers' plans, a weir near Castlekelly diverts the Dodder from its original course (the remainder of the line of the original course, collecting limited inflows, continues on to the southeastern point of the upper reservoir) and into the line of the many-branched **Cot Brook** (*Sruthán an Choite*), from the northern face of Kippure and the saddle between that mountain and Seefingan. This line then takes in (below left) **Slade Brook** (no name in Irish found, possibly just *An Slád*, as a match for the full English name would be "Mountain Stream Brook").

The Slade Brook comes from Seefingan. The full flow is then in turn diverted (shown to left) from the original Slade Brook course (which goes on to the upper reservoir) to a channel, the prosaically-named "*Artificial Watercourse*" - locally often simply "the canal". This runs beside, but separated by a strip of land from, the reservoir on the western side. The Artificial Watercourse flows in the open except for one short stretch of tunnel (upstream entrance shown to left, first arch partly blocked), in turn interrupted by narrow cross-cutting openings (example shown to right), which may be concerned with moderating extreme flows, or simply with holding back unstable neighbouring land.

Small tributaries join along the Watercourse (see left). In case of need, the Dodder can run into the reservoir along the Slade bed (when the water is high, it simply passes over the final weir), and there is also at one point a canal-floor mechanism to recover water from the base of the channel. The municipal water supply offtake runs from the upper reservoir to Ballyboden.

Busy **Glassavullaun (Stream)** (*Glaise an Mhulláin*) (see right), now as with many upper Dodder inflows partly a "tributary" of the water supply system, passes
under the diverted main river and into the upper reservoir, and at least eight small streams also rise on this side of the reservoir, in the Ballymorefinn area, and make their way downhill; some run into the artificial main course, some to the reservoir. Near the northern end of the upper lake a larger artificial stream is routed in from the north west (see left hand side of photo below), bringing most of the volume of two left-bank tributaries, the **Ballinascorney** and **Ballymaice Streams** (merged in the setup shown below), the original lines of which also still join downstream, below the lower reservoir. Just beyond

this the upper reservoir's wide overflow sill can feed much water back to the Dodder. The Ballinascorney Stream comes towards the Dodder main line in a very scenic valley.

Piperstown Stream, from bog and hills to the south and east at Piperstown, with a branching flow from Mont Pelier, and passing between Glassamucky and Corrageen and on to Friarstown (see right). The stream is paralleled, and joined at Corrageen, by a small tributary from Glassamucky. It had an offtake flow under the lower reservoir, into the above-mentioned diversion to the upper lake. However, since World War II, this offtake is diverted, across the top of the lower reservoir's dam, directly into the Ballyboden water supply. This once-substantial stream is known to have driven a corn mill.

At least four short streams enter the upper reservoir on the eastern side, one of these flowing between the old Church of St Ann and the Holy Well of St Ann.

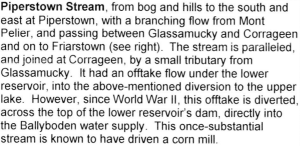

As the river enters the lower reservoir, another small stream joins from the southwest, with at least two left bank flows directly to the lake - and after the reservoir (shown left) is exited, another parallels the river to the west of the roadway, itself west of the river, for a distance before merging in. Beyond, on the right-hand bank, the remnant of the main course of the Piperstown Stream joins, flowing from the area of Friarstown Glen infilled as a tip, to the west of the site of Friarstown House. In the vicinity, the Dodder also receives the residual Ballinascorney Stream on the left bank, and then, just by the entry to the waterworks, the Ballymaice Stream line, also on the left.

Beyond the waterworks, the river cascades at Fort Bridge, at a point called the Sheep Hole

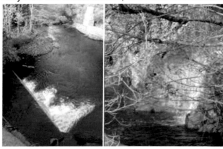

(far left; once used for washing sheep), and a small stream joins on the right, from lower Friarstown Glen (seen near left). The Dodder continues to the north, with some meandering, and the left bank features the remnant of the early stages of the former long Old Bawn millrace, in eastern Kiltipper. The Dodder then bends northeast, and later north again. It cascades impressively under and downstream of Old Bawn Bridge, where steps help to manage the flow (below), and flows on.

In northern Killininny the main line of a mountain stream, the **Oldcourt Stream**, perhaps also **Killininny Stream**, from the Montpelier area, joins. The primary source is on the northwestern slopes of Montpelier, between Mont Pellier House and the Hellfire Club. It runs by Oldcourt, and a tributary joins a little north of Oldcourt Cottages, coming from east of Bohernabreena House, and shortly thereafter another from between Allenton and Oldcourt Cottages, ultimately from branches in the area east of St Anne's Church (sections are visible notably north / northeast of Bohernabreena Cemetery). As the stream line approached the Dodder, near Killakee Walk, it historically split, the main line flowing northwest under Firhouse Road and coming out of culvert for its final reach. The second line ran towards Firhouse (historically sometimes Fur House), crossing Firhouse Road west of the convent and running northwest to reach the Dodder upstream of the City Weir, but is no longer traceable. Two of the classic maps show the main line as the only one, and one shows only the other, albeit merging sooner, but the Ordnance Survey has all branches in detail. As the Dodder continues roughly north, on the left bank are further traces of the large Old Bawn millrace, the tailrace of which passed on to mills south and east of Tallaght village (then a very small settlement on a single street), linked to the Jobstown Stream; it may also have taken water diverted from the Ballymaice and / or Ballinascorney Streams.

At the southeastern edge of greater Tallaght, **Jobstown Stream** (now often **Tallaght Stream**, sometimes **Whitestown Stream**, see left at top of next page) flows into the Dodder after a run of about 8 km. Rising in the hilly area south of Jobstown, and taking in a stream from between Kiltaltown and Lugmore, it crosses northern Killinardan, north of Knockmore and south of the main Killinarden Estate. South of the soccer stadium, **Killinarden Stream** from near Kiltipper, flows in. This stream meanders in the open through the southern "leg" of Sean Walsh Park (left below), passing through a marshy area,

and feeding a lake (see above centre) just before it flows into the Jobstown Stream. The streams in this area have been restructured considerably over the last century. Today, the Jobstown itself carries on east (above right), featuring, with modern ponds, in Sean Walsh Memorial Park, south of Tallaght village centre (both photos below), and going under

Watergate Bridge and running parallel to the Tallaght Bypass (historically it had a winding course, much of this being where the bypass was made).

South of the eastern end of the old village, where Bancroft's Castle once stood, and just south of today's Priorsgate building, a stream flowed in - it may still do so in culvert, but no trace is readily visible. Its line began near the site of Fettercairn House and ran from a point a little to the northeast of Glenanne Park and the original source of the Fettercairn Stream (refer to the Camac section). This line passes under the main element of Tallaght Hospital, and runs south, then southwest under the library and Civic Theatre, then almost northeast. After curving around St Maelrun's Church, it goes southeast, east just north of Main Street, then south towards the Jobstown Stream. Flooding occasionally, the Jobstown Stream once fed the Archbishop of Dublin's ponds, and helped drive mills at Old Bawn and Tallaght.

At Balrothery, north of Firhouse, there is the great City Weir (sometimes "Firhouse Weir" today, view from upstream left top of next page, from downstream right top of next page), at which there was from the 1240's (and perhaps earlier) until recently a major diversion of water to the Poddle, in the artificial channel called the "*City Watercourse*," providing the main water supply of Dublin until the 1700's. While primarily intended to supply drinking water (other flows could be used to source water for washing and small industrial works), the enhanced Poddle also came to be used considerably for industry, to the point that its waters ceased to be potable.

The Rivers and Streams of Dublin (City of Dublin, Fingal, South Dublin, Dún Laoghaire-Rathdown)

A modified version of an early part of the diversion structure is shown below.

This southern City Watercourse was made via western Templeogue, passing the small cemetery there, to meet the Tymon River by Mountdown House and Mills (earlier the Domville Mills) - at a location lying today between Wellington Park and Glendown Grove - and form the "full" Poddle.

The weir has been repaired in modern times, but the sluices and channel of the Watercourse have been less well maintained, and were in part moved and altered to facilitate road development.

Plans were for renovation of part of the City Watercourse, using a line looped back to the Dodder via the Spawell area (as it happens, there was historically at least one "back flow" line from the Watercourse to the Dodder anyway). Part of the line of the Watercourse still exists near Spawell House, as a ditch at the back of a field (behind a service station and playing field) and Sweeney suggests that the line is continued in the county drainage system.

Just downstream of the great weir a small stream from what is now Carrigwood (with a possible intake from today's Carriglea) joins (parts were visible above ground up to the late 1990's, at least). It flowed (and probably still does in culvert) northeast, turning north, northeast and north again within the Monalea development, and running almost parallel to Sally Park. Finally it runs under the roadway in the western part of Mount Carmel Park, and then goes northeast under the adjacent green area.

A little further downstream of the City Weir, under a path leading to parkland just west of the M50 bridge, **Ballycullen Stream** (also known as **Orlagh Stream**), from near Orlagh and Kilakee, joins discreetly (alongside a pipe carrying excess water from the M50). The Ballycullen rises a short distance almost directly north from the Hellfire Club and runs north. Passing to the east of Orlagh House, it takes in a small flow from the west. A streamlet runs partly parallel, joins with other small flows to the east, and merges in halfway between Orlagh House and the cemetery on Gunny Hill. After a crossroads where two more small flows join, the Ballycullen bends, passes St Columkille's Well, from which it may receive water, and swings slightly northwest. By the former hamlet of Ballycullen, it bends to the northeast, picking up small streams, runs west of Woodtown, goes under St Colmkille's Way and on roughly north, and bends past the site of Knocklyon Castle. Finally, it flows along the eastern edge of the former Sally Park demesne, passing Homeville, and runs north to the Dodder. While the upper reaches are largely in the open, the lower parts of the Ballycullen are culverted, especially as it parallels the M50, except for a short stretch east of the Ballycullen Road as it approaches Firhouse Road.

The Rivers and Streams of Dublin (City of Dublin, Fingal, South Dublin, Dún Laoghaire-Rathdown)

Another mountain stream, from at least as far south as Scholarstown (where Woodfield meets Scholarstown Road), trickles in between Lansdowne Park and Woodbrook Park, having been diverted north east (it perhaps used to join the Dodder at the pool known as *Pussy's Leap*, famous for fishing, and reputedly haunted by a pooka, at Knocklyon between Firhouse and Templeogue; the pool was broken somewhat by storm-related flood damage some years ago). The Dodder then comes to the Templeogue Bridge location, historically the site of the first major crossing of the river, and continues to the northwest of Butterfield Avenue, with some notable meandering until it takes a sharp turn to the east after passing under Springfield Avenue, just past Rathfarnham Shopping Centre. Shortly thereafter is the offtake for the upper Bushy Park Pond (hidden in the wooded area), which now also supplies the park's lower pond.

From Rathfarnham to Milltown, three major tributaries flow in, the first with a significant tributary of its own. At the peak of mill activity, and with, additionally, heavy drinking water "take" for Dublin city, there could in the past be little water left in the Dodder channel from Firhouse onwards and even today the river receives a material boost from its strongest tributary, the Owendoher.

The **Owendoher River** (*An Dothra Bheag* – "The Little Dodder" – or *Abhainn Dothair*), which was considered at times as a possible extra city water source, merges in just northwest of Rathfarnham village. It forms from a series of confluences, starting with mergers of several mountain streams.

The prime source line of this active 10-km river, often mapped as Owendoher (upper reaches sometimes as ***Owent(h)rasna (River)***) forms from two stream sets. One is a melding of flows from Killakee Mountain (converging to form **Killakee Stream**), and **Glendoo Stream** from west of Glendoo Mountain, which meet, taking in a flow from Piperstown Gap. The other source line is a multi-branched stream from between the Glendoo peaks, passing Cruagh (also known as **Cruagh Stream**, see right) and receiving two tributaries.

These two, coming from either side of the road from Rockbrook, meet just west of the road at Newtown, near Kilmashogue Cemetery. Notably the more westerly of these supported milling well upstream (at Jamestown and Rockbrook), and at one time, there were two cloth mills and a paper mill near the confluence.

A little downstream, by a school, Woodtown Stream, from south of Mount Venus and Woodtown, joins. This stream, which has several small branches, is partly culverted. The river itself goes on under the high-flying M50 orbital motorway (shown below left) and

continues through Edmondstown, where the remnants of two millraces are still visible.

The growing Owendoher flows on north to Ballyboden (also home to a City Council waterworks), where a very short stream used to join from the west. Passing under Boden Bridge, and then Taylor's Lane, the river bends slightly east of north, and takes in a stream from the southwest at Ballyroan. It then, in the Willbrook area of southern Rathfarnham, merges with the active Whitechurch Stream, its main tributary, coming from the southeast.

Whitechurch Stream (or *Grange Stream*; parts known as *River Glin)* forms from three branches, two from the eastern face of Tibradden Mountain, one from the valley between Tibradden and Kilmashogue. It runs through Kelly's Glen and past Larch Hill, site of Scouting Ireland's main base; it receives a small stream from the old estate, and there was once a swimming pool on its course. The streams flows north by way of Kilmashogue with its bridge. Two small tributaries join from the west, one near the Moravian cemetery in southern Whitechurch and one at the eastern edge of Edmondstown.

The stream continues north through Edmondstown, Whitestown and Harold's Grange, crossing Taylor's Lane and flowing into St Enda's Park. Here it passes through an artificial pond, near which were erected copies of a cairn, an ogham stone and a druids' altar; to the west within this national historical park is the museum to the Pearse brothers, the house formerly known as The Hermitage. The Whitechurch exits the park, passes Sarah Curran Avenue and comes to Willbrook, where it powered a flour mill. It merges into the Owendoher at the meeting of Whitechurch Road and Ballyboden Road. The Whitechurch Stream has a record of rare, but sometimes severe, flooding.

Slightly north of the Whitechurch confluence with the Owendoher a millrace used to run northeast and then east. Collecting the tailrace from the Willbrook flour mill, it swung to the north to fill the mill pond southeast of Rathfarnham Castle, which featured a sawmill.

The tailrace exited northwest, winding parallel with its parent, then cutting northeast, where it used to power another mill, and then flowing northwest of church lands and directly into the Dodder a few hundred metres downstream of the Owendoher.

The Owendoher meanders on north and northwest, passing Rathfarnham Bridge at Butterfield Avenue at one end of Rathfarnham village, then flowing under Springfield Avenue to the Dodder (pictured below – left from beside the Owendoher, right along the line of the Dodder, confluence mid-right).

Before and after meeting the Owendoher, the Dodder runs northeast in its linear park. The confluence occurs just downstream of a footbridge towards Bushy Park (see left) and a little upstream of the "Lipstick Stones" modern foot crossing (on front cover). The Dodder then takes in the *Lakelands Overflow* from the Poddle and Swan systems (shown centre and right, below). The river turns sharply north, and at the city end of Bushy Park bends east, and passes under what used to be called the Big Bridge, with the road directly into Rathfarnham village (now Pearse Bridge). There was once a millrace for a sawmill and printing works, starting at a weir ¼ kilometre downstream of the bridge and paralleling the Dodder course until just upstream of Orwell Bridge; traces can be seen today, including at

the mentioned weir (see above left). The Dodder runs directly east until just north of the Church of Ireland Theological Institute and Library, northwest of which it receives the Little Dargle.

The **Little Dargle River** (occasionally ***Badger(s) Glen Stream***, ***Glen River*** or ***Sohlang River***) rises in a small wooded area north of the "Fairy Castle" part of Ticknock, where the northern slopes of Two Rock Mountain meet the western slopes of Three Rock Mountain, and flows for about 8½ km on its main line. It starts by heading roughly north, collecting mountain branches, then runs under the M50 to pass under College Road and enter Marlay Park, there branching and feeding ponds. In the southern reaches of the park, it meets one tributary, from northern Kilmashogue, and another joins near the main lake, coming from the southern edge of the park, and College Road. The Little Dargle then heads north (see above), exits the park and runs in culvert to Ballinteer, coming out from piping in green space to receive a stream that runs, mostly openly, from the Ticknock Road area.

The Ticknock tributary forms from several close branches, one rising at a spring on the side of Ticknock Road, one in a field a little to the southwest, and one a little northwest of the holy well called Grumley's Well. Waters from the latter well have contributed to and may form the source of the latter branch. The Ticknock tributary passes Kingston Hall and Park, Ballintyre Grove and goes on under Heather Lawn and Stonemason's Way.

The "boosted" Little Dargle winds through the aforementioned green area and passes the place where the Phibb's Weir overflow to the River Slang - later taking in the natural Wyckham Stream tributary of the Slang - operated, once boosting the Slang's flow to the mill at Dundrum. It goes on by Nutgrove Avenue (this part used to have twin courses) and then at the north end of Castle Avenue takes in a tributary (possibly **Churchtown Stream**) from two branches, one from beyond Nutgrove Avenue south of Oakdown Road, and one from Beaumont Avenue and possibly north of Weston Road. Later it also meets a flow from the Braemor Road area and the site of the grounds of the historic house once called *Landscape*.

Finally **Castle Stream** joins the Little Dargle; rising in Rathfarnham, this passes Nutgrove Avenue, and enters the grounds of Castle Golf Club, where it is met by a stream from the Loreto Abbey grounds (sometimes **Abbey Stream**), and an overflow streamlet from the Rathfarnham Castle grounds. This latter used to come southeast from a long narrow fish pond south east of the castle and west of the castle pond (which was in fact a mill pond) but now seems to run from the castle pond itself. After twisting and turning near what is now Woodside Drive, the streamlet meets the river near the "triumphal arch" (Lord Ely's gate to Rathfarnham Castle lands), and the combined flow runs the short further stretch in culvert to the Dodder opposite Orwell Park, a little to the west of Mount Carmel Hospital (see right).

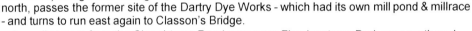

The Dodder then continues past Orwell (or Waldron's) Bridge and having run east and north, passes the former site of the Dartry Dye Works - which had its own mill pond & millrace - and turns to run east again to Classon's Bridge.

A small stream from the Churchtown Road area near Flemingstown Park runs north and loops out to the west, turning back east behind Green Park, near the rear of the Russian Embassy grounds, then runs roughly north with Orwell Road, and on in that line after the road turns west-northwest. Linked to ditches from the roadside boundary of the Milltown Golf Club grounds, it reaches the Dodder at a weir between Dartry Park and Classon's Bridge. One source suggests that another watercourse, from the vicinity of the northwestern part of the grounds of St Luke's Hospital, around Highfield Park, used to come to the Dodder just upstream of Classon's Bridge, but this cannot be traced today. Downstream of Classon's Bridge, the Dodder goes east, under the Luas line (on the famous Nine Arches viaduct), and then bends to run northeast past Milltown.

The **River Slang** (*Abhainn na Stéille*) (or **Dundrum River** or **Dundrum Slang**), the upper portion of which (to the meeting with Wyckham Stream) is sometimes **Ticknock Stream**. The Slang rises on Three Rock Mountain, flows through Ticknock, and runs north, cascading down steps to an access road and going under the M50 motorway at Junction 13, before continuing through Ballinteer, where historic mapping suggests an eastern tributary used to join. Where Wyckham Way bends northeast, the Slang follows the line of the old Ballinteer Road (above left), and is joined from the west by the Wyckham Stream. **Wyckham Stream** is shown on old maps as a short flow going directly east to the Slang. At some time in the 19th century it was extended westwards with an offtake from the Little

Dargle, to boost mill water supplies. It receives a tiny stream from the southwest, and occasionally floods by Acorn Road.

The Slang then makes a wide turn to the east, going under Wyckham Way as it loops east to Dun Emer Drive and north almost to the Balally Luas stop. Receiving a small eastern tributary, the Slang flows back west via an apartment complex (below left) and the

Dundrum Town Centre shopping facility. The main line of the stream crosses to the western side of the mall, while the historic millrace comes to the restored mill pond (the former Manor Mill site, once hosting perhaps Ireland's largest laundry, has been remade in turn as the Pye factory and now for retail), a feature at the Town Centre (above right).

The Slang then runs roughly north past Dundrum village to Windy Arbour, which held a millrace. Early mapping suggests that two western tributaries used to join in Dundrum, one at the southern end of the village, and one just south of St Nahi's (Taney) Church; these are not shown on later maps, and are not now traceable. Going just west of Dundrum Road, and passing Frankfort and behind Millmount Grove, the Slang goes into culvert between Dundrum Road and Patrick Doyle Road, and reaches the Dodder south of Milltown, a little downstream of the Nine Arches Bridge (now used by the Luas) and opposite Alexandra College, after a main channel run of about 8 kilometres. It floods in the Dundrum area at times - one study shows more flood points on this little river than on any other part of the Dodder system. "The Slang Greenway", a walking route promoted by local authorities, links Marlay Park and Dundrum along the combination of part of the Slang, the extended Wyckham Stream line, and the Little Dargle.

The Dodder winds on roughly northeast and then, after Clonskeagh Bridge, almost north until the major weir beside Beech Hill Road, after which it runs northeast to Angelsey or Anglesea Bridge. Maps suggest that a small tributary joined downstream of Milltown Bridge, northeast of Alexandra College, coming from the Richview Park area, and another on the left bank at Clonskeagh. There may also have been a streamlet just upstream of Anglesey Bridge, from the vicinity of the bus depot (formerly a tram depot).

Milltown and Clonskeagh featured extensive mill works in times gone by - there were millraces both south and north of the river for ironworks, for example, with the northern one becoming part of the Clonskeagh Smurfit works. Just downstream of the latter and the small stream mentioned in the previous paragraph, the longest Dodder millrace began, running northeast, across Eglinton Road and beyond, supplying the Magdalen Laundry at Donnybrook, a flour mill, and the Johnston, Mooney and O'Brien bakery complex, and rejoining the river near Balls Bridge. In even earlier times, this millrace carried on to Irishtown, joining the Swan River just before it merged into the Dodder.

Downstream of Angelsey Bridge and the beginning of Donnybrook, the Dodder goes north, a little northwest, and then resumes a northeasterly line.

The small, now-hidden, **Muckross Stream**, comes from between Rathgar, Rathmines and Milltown, around Cowper Road and Temple Gardens, in the area once known as "The Bloody Fields," part of the broad area known for an attack by native Irish on citizens of

Bristol-controlled Dublin in 1209. Passing through a part of the Jesuit lands at Milltown (Milltown Park and Bewley Park, within which are the Milltown Institute, Gonzaga College, the Jesuit Provincialiate and other institutions), and running parallel to, and northwest of, Hollybank Avenue, near Sandford School, it crosses Sandford Road to the Muckross Park College campus (the original Muckross Park House may have been named after it). It then passes Belmont Avenue, Donnybrook Manor and Belmont Park, still in culvert, and travelling near Donnybrook Fire Station and Garda Station, crosses under Donnybrook Road to the Pembroke Cottages area. Here, on the western side of today's Donnybrook Mews, the Muckross used to be channelled under the last-mentioned Dodder millrace. Its line then passes under Eglinton Terrace at the point where it bends, and winds roughly parallel with that street, crossing the northern extremities of the rugby grounds, coming to the river opposite the midpoint of the Hazeldene development.

Facing the Royal Dublin Society (RDS) grounds across Anglesea Road, the Dodder runs north, passing the former Pembroke Town Hall and Ball's Bridge, which gave its name to the whole area. The river is tidal from the weir at Ballsbridge. It heads northeast to New Bridge, where it turns to flow west of north towards the Liffey, just southeast of the Lansdowne Road stadium grounds, and northwest of the hist site of Haig's Distillery. Permission was given around A.D. 1300 for a supplementary supply to be taken from this vicinity to the mill on the River Stein in front of All Hallows Priory (later the site of Trinity College), in a project not dissimilar from the Dodder supply as used to boost the Poddle - but it is not clear if a canal or pipe was ever built to carry out this plan.

The last Dodder tributary, the **Swan River** (or **Swan Water**), whose many branches total over 15 km (some say 17 km) of channel, flows through Terenure, Rathmines, Ranelagh, Ballsbridge and beyond. It has been greatly altered, and much of its system integrated into city sewers. The Swan begins in the Wainsfort area and crosses Kimmage Road; just south of Mount Argus, this early stage is largely diverted to the Poddle.

The main course continues under Kenilworth Park and across Harold's Cross Road, to take in a tributary. The line of this branch, with an offtake supply from the Poddle (the "*Lakelands Overflow*", built in 1937 to relieve pressure on the Poddle line, replacing some ditches which had formed a link up to 1930), runs to the lake at Terenure College. The lake has an overflow to Bushy Park and the Dodder (see the Dodder section) but the line of the other flow from the lake, the **Olney Stream**, continues east along the southern edge of what was the City of Dublin VEC's southern sports grounds, and across Terenure to Rathmines, past the sites of a former lake and quarry on the stream line, at Terenure Road North (it is not clear if this stream still has a regular flow today). Sweeney labels this branch as a whole the "Roundtown Stream" (from an old name for Terenure).

The main Swan line bends southeast from Harold's Cross, and northeast west of and parallel to Rathmines Road (crossing the site of a modern municipal sports centre), then southeast again, taking in a small tributary coming east and southeast from Harold's Cross via the Cathal Brugha Barracks area (Sweeney calls this the "Blackberry Brook"), and with an overflow connection to the Grand Canal Tunnel sewer tunnel to the north.

Beyond Mount Pleasant Square, the little river's line heads east through Rathgar and passes through Swan Park, where it probably supplies the duck pond (exactly on its historic line). It then runs under the Swansey Terrace part of Chelmsford Avenue and goes on just south of the junction of Chelmsford Road and Sallymount Avenue.

At Chelmsford Road the Swan line meets a small branch from around Garville Avenue via Belgrave Square, which Sweeney labels "the Bloody Fields Water" from a name for the broad area remembering the attack by native Irish on citizens Dublin (as mentioned in the Muckross Stream section). At Sallymount Avenue there is another overflow line to the Grand Canal Tunnel Sewer, heading north-northwest along Leeson Park.

The Swan reaches Donnybrook between The Appian Way and Bloomfield Avenue, going along the northern boundary of the historic Royal Hospital / Bloomfield Hospital grounds, and taking in a small southern stream from Moyne Road. It passes just south of Swan Place and crosses Morehampton Road. The river then runs in culvert along the southern boundary of the An Taisce (Ireland's National Trust) property "The Grove" - a stretch of river here, south of Wellington Place, was one of the last unculverted parts of the Swan. The Swan historically wound across the northern parts of Herbert Park, with a small channel heading for the Dodder (near the former Johnson, Mooney & O'Brien bakery complex) - this channel was locally called "the Swanee." A new main course now divides from this near-dry line at the western end of Clyde Road, continuing northeast.

This modern river line passes several embassies on Clyde Road, goes through Church of Ireland grounds at the junction with Elgin Road, and under properties just northwest of the landmark Embassy of the USA, all in the area once called Baggotrath. By Pembroke Lane it takes in one last tributary flow (in culvert), from south of Portobello Bridge via Ranelagh, Lesson St Upper and Waterloo Lane, which Sweeney tags as "Baggotrath Brook."

The Swan line runs citywards along Pembroke Road (once Hammersmith Place) and then abruptly bends to flow along the frontage of the Ballsbridge Hotel. It crosses Lansdowne Road (once "Watery Lane") and later Lansdowne Park, turning to flow east just south of the Beggarsbush Barracks site, and then going south of Havelock Square and north of the Aviva (formerly Lansdowne Road) stadium. Here it once turned northeast and widened, joining the Dodder just upstream of London Bridge, by Bath Avenue. The long millrace mentioned earlier joined the last stages of the Swan also.

Today what remains of the Swan Water flows almost straight along the stadium boundary line - part of its final stretch was further realigned during work on the new stadium - and can be seen outfalling to the Dodder from the path on the other side (above left). Historically the Dodder mouth, separate from the Liffey mouth, was a wide expanse of water extending far south of the current confluence; then, Ringsend and Irishtown were rather isolated from the city, and the Swan Water flowed directly into this Dodder estuary.

Now, the Dodder flows on beneath London Bridge, between stone walls (above right), passes Shelbourne Park greyhound racing track and goes under Ringsend Bridge, to reach the Liffey on the approach to Ringsend, shortly before the East Link Bridge. At one time,

there was even a ferry crossing across the Dodder mouth here, opposite a former freight rail depot, now the Three Arena (the former Point Depot and O2 venue). The Dodder-Liffey confluence is combined with the end of the Grand Canal, the sea locks of which allow boats to join the final few metres of the Dodder, just below the Grand Canal Docks, and the canal basin which holds the Waterways Visitor Centre (see left above).

The Rivers and Streams of Dublin (City of Dublin, Fingal, South Dublin, Dún Laoghaire-Rathdown)

Rivers and Streams of Southern Dublin

Note: Some southern waters are hard to fully trace - many are heavily culverted and contemporary maps and observation must be aided by historical drainage maps, etc. Most were substantially visible into the 1960s. A further issue is that the naming and structure of the leading watercourse system, that of the Shanganagh River (parts known as Loughlinstown River), is not always straightforward or consistent.

The watercourses, north to south by mouth, and then those flowing into County Wicklow, are:

- **Elm Park Stream**, main branch rising between Taney Park and Rosemount Estate near Dundrum and Goatstown, joined early on by a tributary from Rosemount Park, and proceeding roughly north. Passing between the Central Mental Hospital and the Our Lady's Grove complex of convent, schools and latterly housing, it flows by Mulvey Park and the Gledswood roads, and White Oaks, in Friar(s)land (see right), and crosses Roebuck Road. It runs by Clonskeagh Mosque and turns to go east, receiving a short tributary from another part of Friarsland, while passing the French and German international schools and UCD's horticultural garden. It travels on through the main University College Dublin campus, known as Belfield  after one of the former local "big houses", and takes in a short tributary (to left) near the Arts, Commerce and Law building, and UCD's archives, before going into culvert. It travels on northeast, past the campus's central pond, and in the northeastern corner of Belfield, in the townland of Priesthouse near the county boundary, merges with an active tributary from the Richview area of Clonskeagh. Nowadays the confluence appears to actually happen under Stillorgan Road, north of Belfield's main road access bridge-and-underpass combination.

The Richview tributary rises near UCD's Planning and Environmental Policy Unit, and takes in a further tributary from between Belfield Close and Wynnsward Drive south of the University Lodge. It flows under Bianconi Bridge (shown right) and through a wooded area of the campus, with several ponds (see below left), before going into culvert as it nears

Stillorgan Road. The Elm Park emerges within the golf course of the same name, running northeast, east

and northeast, and entering the former Merrion Castle grounds, south of St Vincent's Hospital. Here it runs in the open (see both photos at top of next page), north of several apartment buildings and south of a number of care facilities, before diving under Merrion Road, a service station and the railway line.

It comes to the South Bull sands just southeast of *Merrion Gates* level crossing. At the outlet, there is a wide sand trap, from which the stream's "estuary" runs; when the tide is out, it wanders across the sands towards Poolbeg (photos below).

- **Trimleston Stream** (or *St Helen's Stream*) from south of Foster's Avenue, that - having run by the edge of UCD's main campus and crossed under the Stillorgan Road dual carriageway - passes through the former St Helen's estate (site of the Irish Christian Brothers Novitiate for decades), now with housing beside a hotel; the stream runs a little east of north (with one short open stretch, to right). It turns to flow west south of the return end of St Helen's Road, and then, passing a pond, bends sharply to the north. It runs between the western line of St Helen's Road and Trimleston Gardens, and at the northern end of the latter passes the former site of a pool on a small tributary from a spring under what is now Merrion Crescent.

The Trimleston comes to the South Bull by way of a blocky culvert along the northern edge of Booterstown Marsh, near the boundary between Dublin City and Dún Laoghaire-Rathdown (the stream mouth is just outside the city boundary). The stream does not today materially interact with the Marsh, or the passing Nutley Stream (see next section), though there can be some seepage to the marsh and one source says that a greater flow can be released by valve or sluice if the wetland becomes too dry. Crossing under the railway, the stream emerges to run northeast across a modest patch of dry sand, which has begun to form small dunes, and on to the sea (at low tide it travels across the sand and mud for a considerable distance).

Historically, the Trimleston may have been a shorter stream, beginning at a hollow south of St Helen's Road, and later connected, by extending a ditch to the south, with another line - but this is not certain. After the railway was built, the Trimleston was "captured" by the Nutley culvert, but a separate outfall was made after the severe flooding of 1963 (when the old Nutley line, from Ailesbury Park to St Alban's, reverted to "lake", with Ailesbury Park and

St Alban's under several feet of water for more than 40 hours, and the Merrion Gates level crossing under up to 18 inches - this episode, featuring Dublin's worst watercourse-led flooding for many years, also impacted Goatstown, Clonskeagh and other areas).

One of the southern watercourses nearest the city was known historically as **Clarade**; it is probable that this was either Elm Park or Trimleston Stream but it could also have been a reference to the Dodder.

- **Nutley Stream**, a much-modified watercourse, rises between Richview and Beech Hill in Clonskeagh, and runs roughly east in culvert, by Airfield Park, to RTE's Montrose lands, from where most of its flow is diverted to the lower Elm Park Stream (see above); this link was made to manage flood risk as the watershed of the middle Nutley was developed, with large areas concreted over. It takes a small tributary from south of Ailesbury Road, behind the French Ambassador's residence, possibly **Ailesbury Stream**, just west of the Merrion Centre retail and office complex. It crosses Nutley Lane, then Merrion Road, and then goes along the southern edge of the gardens of Ailesbury Park. It turns southeast in a 19th century railway diversion (prior to this, the stream passed, via a sizeable lake covering much of southern Sandymount, to the sea at Sandymount Martello Tower; the lake area is now developed as housing – notably what is now St Alban's Park – with pumps working below ground to keep it dry). Passing Merrion Gates to the west, the Nutley, in culvert, goes *over* Elm Park Stream, and turns east, coming into the open in a ditch alongside the inland face of the railway embankment behind an old dye works site (now an office building). It then runs along the eastern edge of, and feeds freshwater to, Booterstown Marsh, the only remaining saltmarsh area in southern Dublin Bay, and passes under the land link from car park to station at Booterstown DART stop.

In the section of ditch south of marsh and station, sometimes called "Williamstown Lagoon" or "Williamstown Creek," small surface water drains add a little to the flow. The stream finally turns to the sea via a tunnel through the rail embankment (see below left) a little down the coast from Booterstown Station, beyond the Martello Tower in Williamstown (the small district between Booterstown and Blackrock, also home to Blackrock and Willow Park Schools). This underpass is the last remaining of three such tunnels which drained a once extensive marshy area, and is within the administration of Dún Laoghaire-Rathdown. The rising tide feeds saltwater back up to Booterstown Marsh, ensuring it remains brackish (part-saltwater, part-freshwater), allowing it to host an interesting range of plants (it is also very popular with birds). The Nutley continues across the mudflats at low tide, but is fully taken into the sea at higher tide (visible as a current in the water below right).

There are a couple of references to a "Booterstown Stream" but no historical flow of that name could be clearly identified - and one set of references, when traced on a map, in fact related to a drainage line "back" from Williamstown towards the Elm Park Stream (this line was a key element in the great flood of 1963). Previously, the whole coastal area from Merrion to Blackrock, inland from the railway line, could flood tidally, and contained several semi-permanent stream lines.

We now fully depart Dublin City, and leave municipal sources, including Sweeney's *Rivers of Dublin,* but the *Hidden Streams* volume comes into its own with many of the following waters mentioned, though with limited details on the courses of the waters.

- **Priory Stream**, today beginning in the area of the former Stillorgan Demesne between Priory Avenue and Stillorgan Park. It runs slightly west of north, crooks sharply east to the north of Priory Avenue, then turns north again to cross Waltham Avenue going east. It used to run through the grounds of the controversially demolished Frescati House (occasional mentions as *Frescati Stream* occur), and still has short stretches in the open in that vicinity, before crossing under the road to the southern edge of Blackrock Park (just north of Deepwell House and gardens) in a small deep valley (see below left), which levels off into a tidally-influenced channel (below centre). The stream's mouth was historically at the north eastern edge of Blackrock Park, which was itself developed from swamp and an older "walking garden" which had two culverts to the sea cutting through the embankment. It was moved to the second culvert, at the mid-point of the park, with a basin at the sea outfall point, in modern times. The stream does not directly supply the Blackrock Park pond, which in fact holds rather salty water, mostly being intruding back-flow from the Priory's sea outlet basin (below right). In the late 2000s a proposal was made to remodel the park, including diverting

the stream to flow to the pond mid-way through the park, and forming an integrated constructed wetland area (see the Finglaswood Stream part of the Tolka section for a note on a working example of the ICW approach). The Priory Stream has at least one tributary, in the lower reaches, possibly sometimes *Mount Merrion Stream* after its source area.

Priory Stream seems to be a remnant or continuation of what was sometimes *Kilmacud Stream* ("appears" as documented by usually thorough Ordnance Survey operatives, and certainly appearing to fit the contours - but one historic map shows the Kilmacud flow as historically linked to Glaslower, and the Priory flow as separate). Today this is linked to the Carysfort-Maretimo Stream (see just below) but is primarily integrated into a long drainage line connecting the Sandyford business parks to the Kill o'the Grange Stream. It rose originally in Ballally, in marshy land near what is now Blackthorn Grove, and flowed north, passing under the Harcourt Street railway line, and on to and through the grounds of a house called "Lakelands" in Kilmacud, feeding a pond. It went on north, received a small western flow and turned to flow east just south of Kilmacud Road. The stream ran down Stillorgan Hill at an angle to the south of Kilmacud Road (the line would cross along the northern edge of today's GAA grounds); now there is a deep culvert on the hill. Crossing the road, it ran northeast, then turned to the north near Stillorgan House. From here on, the modern Priory Stream runs.

- **Carysfort-Maretimo Stream** (historically *Glaslower*, sometimes *Leper's Stream* or *Leopardstown Stream*, from a site linked to the old Leprosy Hospital of St Stephen) is an active watercourse, with some history of flooding, flowing from the slopes of Three Rock Mountain just above Sandyford village, via later-built housing estates, and under the Sandyford business parks to the Sandyford Reservoir complex (sometimes it is *Sandyford Stream* around here). After going partway around the reservoirs, the stream comes to Brewery Road and enters a culvert. It then runs northwest, going under the grounds at the Leopardstown Inn, and returns to the open to pass the former brewery site, where its waters were used (sometimes this part is *Brewery Stream*), north of the former Esso Ireland HQ, now apartments. Running in culvert, partly under the western carriageway and central median of the Stillorgan road, it passes the site of Darley's Well (at Merville Road) and then turns northeast (it used to supply a bathing pond here, east of the graveyard) and runs along the south side of Stillorgan Grove. Somewhere between Brewery Road and Stillorgan Grove, there is an overflow link to the Kilmacud Stream and the drainage line of which it is part. The Carysfort-Maretimo goes northeast under the grassy area at Orpen Green, then north to Stillorgan Park and then northeast in the Avoca Park / The Cloisters area. It passes the Carysfort lands, where it forms Carysfort Pond just south of the former Carysfort College (once one of Ireland's main teacher training colleges, and now UCD's Graduate School of Business). At the eastern end of the pond, the original stream line runs east south of Carysfort Hall, to Avondale Lawn, with short culverted sections, while a fan-shaped weir leads to a relief line (there will be twin lines when all improvement works are complete) across Carysfort Park. (Views from the Carysfort / pond reaches below).

The lines rejoin east of Brookfield Terrace and go northeast along the edge of the sports ground behind Brookfield, where there is a footbridge. After forming a small pond, the stream flows on to the east of Barclay Court before crossing Temple Hill, then turning sharply north and running in culvert along Temple Road. Running near Blackrock's Roman Catholic church, to the east of the parish centre, in a deep but open cutting, it goes under Newtown Avenue, passes Sweetman House and runs down to the sea out of general sight in the rocky area between Idrone Terrace and Maretimo Gardens West, exiting to a rocky outfall through a railway embankment cutting. The Glaslower or Carysfort-Maretimo Stream has a record of flooding at several locations and is the subject of flow-improving and anti-flood works since the late 2000s.

- **Rochestown Stream** (or ***Mickey Brien's***, ***Carrickbrennan***, ***Ballinclea*** or ***Monkstown Stream***) comes, in culvert, from eastern Rochestown, around Ballinclea. Its line begins either on Ballinclea Road or by today's Roxboro Close, and it runs roughly east and northeast from Ballinclea Road across Avondale and Thomastown Roads, to parallel Beechwood Lawn, and later Pearse Park, where it turns north. It exits Sallynoggin in line with O'Rourke Park and goes northeast, in the open, across the southern part of the former Dún Laoghaire Golf Club course. It turns sharply north near Ardmore Park, and may here meet a small tributary from on or near Dún Laoghaire Institute of Art, Design & Technology lands. It meanders to Kill Avenue, then crosses Monkstown, roughly parallel to Fitzgerald Park and Meadowlands, then tending northwest, to the former site of a large marshy pond, east of the ruins of Monkstown Castle. Passing under Mounttown Road, it runs between the small graveyard and the former Monkstown Park, then flows along the eastern end of the gardens of the part of Carrickbrennan Road approaching Pakenham Road, going into culvert, and preparing to receive the western Stradbrook Stream tributary.

Stradbrook Stream (possibly ***Strad Brook***; also ***Carrickbrennan Stream*** and ***Monkstown Stream*** in at least one source each), rises in two branches, both now in culvert. The line of the western branch, with a more natural form, begins in what is now green space west of Rockford Park, $^1/_3$ km north of Deansgrange Cemetery, and heads north and east near

Rockville Crescent, and north again, just east of Deansgrange Road, from just before the junction with Stradbrook Road. The eastern branch line runs from the former edge of Newtownpark House lands, now the inner end of Brookville Park, in a series of straight lines, northeast, then northwest north of Rowanbryn, then northeast again, across parts of Rockville, and east and southeast across part of Brooklawn, to join the first branch as it begins its northern run. Where Stradbrook Road passes by Gleann na Smol, the stream turns east, then northeast towards Monkstown Road, which it roughly parallels east and southeast, flowing in the open south of Drayton and Heathfield, and north of the Cheshire Home, and Alma Park (see right). It runs south of St Patrick's Church and back into culvert, follows the property line north and east, crosses Carrickbrennan Road, and runs in culvert along the boundary north of the Friends' Meeting House, shortly after joining the main stream.

The unified stream flows towards Old Dunleary, going northwest of the former Monkstown Hospital and adjacent named well. Following infilling of the old harbour, and development of the modern Dún Laoghaire Harbour, it now comes to the sea at the West Pier. It causes occasional floods as it falls sharply seawards, notably near the Purty Kitchen venue.

- **Glasthule** (*O'Toole's Stream*), rising in the Ballinclea area by Rochestown, below Rocheshill. The main branch flows roughly north, through parts of Rochestown and Glenageary, and eventually Glasthule, to reach the sea midway along the coast at Scotsman's Bay (sometimes historically Scotch Bay; outfall shown left), to the east of the town of Dún Laoghaire. Buried many years ago, its culvert's final stretch follows the line of the Link Road, and its outfall is discreet but distinct (the seven or so sewer outfalls in the bay were diverted to Dún Laoghaire's West Pier processing facility many years ago). While some early maps show a single stream, others show at least one eastern tributary, possibly ***Thomastown Stream***, from east of the house known as Bellevue in Thomastown. This runs north and then west, and another comes east and south east from Monkstown, with all joining just south of the Glasthule Stream's crossing of "the Metals" - the line of the old stone-hauling railway from Dalkey Quarry to the Harbour. This stream has been fully culverted for decades.

- **Kill o'the Grange Stream** (or ***Deansgrange*** or ***Dean's Grange Stream*** or ***Clonkeen Stream***, pictures on next page), rising near the church at Kill Avenue, by Clonkeen Road, at the central point of the Kill o'the Grange area (historical maps show a start a little to the northwest of Kill Lane).

The stream cuts diagonally across the lands between Clonkeen and Pottery Roads, then runs under Johnstown Road. It is joined, about two thirds of the way from Kill o'the Grange to Johnstown Road, by a western line (some say properly *Deansgrange Stream*), from by Kill Lane in Deansgrange (it passed the house called "Deans Grange"), more than halfway from Deansgrange Road to Stillorgan Road, and bisecting the land between those roads, before running under Clonkeen Road. The stream crosses eastern Cabinteely and goes into

Kilbogget Park. Early in the park, between Shrewsbury Wood and Doonanore Park, are man-made ponds, after which the stream goes underground, reemerging around Wyattville Road and winding south and southeast, past roads including Glenavon Park and Glencar Lawn, then heading east to the north of Achill Road, and on. It comes to the sea across the stones of Killiney Strand (see below: downstream left, looking back on right),

a little to the north of the mouth of the Shanganagh. The main stream is open for a majority of its course, much of it canalised, some in concreted channel. For several decades, the head of the stream has been connected to the previously mentioned diversion drainage line which extends from the Sandyford business parks area, taking in the Kilmacud Stream, flowing north, east to Stillorgan, and southeast, and also gathering in some water from the Glaslower (also known as the Carysfort-Maretimo Stream).
- **Shanganagh River**, its system the most extensive in Dún Laoghaire-Rathdown, forming in a series of mergers: St Bride's / Cabinteely / Foxrock Stream with Carrickmines River (from Ballyogan Stream and tributaries, and Racecourse Stream) to become what is sometimes called the Loughlinstown River (sometimes specified as the north or Carrickmines branch), flowing between Cherrywood and Loughlinstown, and merging with Bride's Glen Stream (sometimes known as Loughlinstown River (south or main branch)), the unified flow coming to the sea at Killiney Strand / Killiney Bay within Shanganagh, a little south of the mouth of the Kill o'the Grange Stream and with a rather larger water volume. At times, notably on some maps, the appellation *Loughlinstown River* is applied more widely within the system. The system's elements lie in an arc from Foxrock southwest to around Kilternan, and pass through a mix of undeveloped area and densely-populated suburbs.
  St Bride's Stream (or *Cabinteely* or *Foxrock Stream*) rises north of the western end of Foxrock's Westminster Road, then turns east. The longest branch stretches from deep inside Foxrock Golf Club's grounds, and is mapped at least back into the 19th century; it has been buried for at least many decades. It meets with lines mapped for centuries which now come from between Hainault Road and Plunket Avenue, all these and a short line from the northwest meeting a little to the south of Westminster Road. Near the old site of

Foxrock Post Office, the stream runs northeast and southeast, going through the former grounds of Kilteragh, once home to model farm elements (Horace Plunkett, co-founder of the critical agricultural co-operative movement in Ireland, and in effect founder of the Irish Department of Agriculture, lived there before the house was burned out). The upper reaches appear to be entirely in culvert today.

Just south of Gordon Avenue, the stream turns roughly north of east (the line is visible in property boundaries); it goes on to traverse Cornelscourt. The St Bride's flows north of Kerrymount Close and south of Kerrymount Green, surfacing by Glen Lawn Drive to go southeast through the Glen Lawn residential development and on to feature in Cabinteely Park (it bisects the public park in a shallow valley, with a pond area; other water channels in the park probably also contribute to its flow). What is here sometimes the Cabinteely Stream continues roughly south, passing through western Brennanstown, and meets in Lehaunstown / northern Loughlinstown – just west of the end of Beech Park – with the Carrickmines River, to form the **Loughlinstown River (north or Carrickmines branch)**.

Carrickmines River, as mentioned, forms in the meeting of the Racecourse Stream and Ballyogan Stream at Carrickmines (just west of the second Carrickmines roundabout, near a northern section of Glenamuck Road and Ballyogan Grove) and flows on broadly east, with multiple small cascades.

Racecourse Stream, a feature of Leopardstown though no trace of naming as "Leopardstown Stream" has been found, rises in the area between Murphystown and Leopardstown Roads, and can be seen going under the Luas line before passing below the M50 motorway to run south of Leopardstown Hospital and along the southwestern edge of Leopardstown Racecourse (it can be seen in this vicinity from the M50).

Ballyogan Stream (sometimes *Kilgobbin Stream*) comes from the northern slopes of Three Rock Mountain. One branch begins at a spring in the area of Barnaculla, with a well site nearby, and runs north to Enniskerry Road, then north of east to meet the other branch. That second branch forms from streams from the slopes of Three Rock south of Ballyedmonduff Road, with the combined flow running north and through Fernhill Gardens (private ornamental gardens on show for many years, closed to the public for some time; a glimpse of this tributary is shown to the left).

The two branches merge just west of the Belarmine housing development and the Ballyogan goes north and then turns east to pass below Kilgobbin. There were three historic tributaries from Kilgobbin, one of which may have been *Kilgobbin Stream*. The first of these goes southeast towards Kilgobbin churchyard, and then straight north (the latter part at least still seems to flow), while the second and third form from a stream rising in branches from fields south of Stepaside, which splits, one line passing Kilgobbin village and its castle ruins, while the other swings out east and then runs north and northwest. The Ballyogan flows on southeast, taking in a small left bank tributary and then a larger one from branches in the Jamestown area. The stream then runs by the former Ballyogan tiphead in dual culverts, which merge north of the retail park called The Park in southern Carrickmines. The twin lines have existed for many years, though early mapping shows only the southern, winding one, so it is possible the more direct northern line was made with the idea of driving a mill.

Historically the Ballyogan Stream turned north just down-stream to meet the Racecourse Stream and form the Carrickmines River, which was then joined further downstream by two distributaries of the Glenamuck Stream, but this was changed when the M50 motorway was extended to this area. Today, **Glenamuck Stream** joins the Ballyogan by the Glenamuck roundabout in southern Carrickmines. This stream comes from branches from west and northwest of Golden Ball, which merge into two lines, and then a single stream just west of Glenamuck Road, which receives the **Golf Stream** (from a slightly complex set of channels) and also takes in a small tributary from the Cairnbrook development east of the road and other streamlets, passing The Park retail development to come to the Ballyogan Stream.

Below Carrickmines, a streamlet joins on the left, then a larger stream on the right, within Lehaunstown or Laughanstown, the latter coming under the M50 from near Carrickmines Golf Course), and the little river passes through a small wooded area at Brennanstown. It then meets the Bride's Glen Stream, also known as the main branch of the Loughlinstown River, southeast of the historic centre of Loughlinstown and just east of the N11.

Bride's Glen Stream (also known as the *Loughlinstown River (south or main branch)* and not to be confused with the St Bride's / Cabinteely / Foxrock Stream, discussed above) flowing broadly east from Three Rock Mountain, and often taken as the prime flow of the Shanganagh system. It historically rose in and around, and channels still run from, a wooded area of western Ballyedmonduff west of Vard's Lane, south of the peak of Three Rock Mountain and northeast of Two Rock Mountain. The line runs roughly east and northeast (north of Taylor's Folly), meeting a streamlet from a spring by Ballyedmonduff Road and possibly another small tributary as it bends to run southeast passing the portal tomb at Kilternan. Northwest of the site of Kilternan Abbey, the stream receives some eastbound flows, one of which comes from a broad wooded area well above and to the west of Ballyedmonduff Road, and runs northeast, east and southeast, collecting smaller channels. It continues towards Kilternan, receiving streamlets from the Verney area, and then heads east to the south of Ballycorus Road, skirting the Lansdowne / Old Wesley out-of-town rugby grounds. At the site of the former Ballycorus Lead Works small flows and an offtake from the main stream fed a mill pond (the mill system no longer operates) and then the stream heads north across the road (bridge parapets can be seen just before the Mine Hill Lane turnoff) and goes on east. It passes Heronford Lane, named, unsurprisingly, for a former ford, and then parallels Bride's Glen Road to the north (see below left), before cutting over to the south

of the road just before it passes under the M50 motorway (above right). Shortly afterwards, it

flows under Cherrywood Bridge (passing the Dublin fresh water main line from the Vartry complex), going under Bride's Glen Road twice more, and heads almost north alongside Cherrywood Road. It then passes below the N11, and just east of that road meets the northern branch Loughlinstown River at the northeastern end of Loughlinstown Woods, forming the Shanganagh River.

The Shanganagh runs southeast and east through Loughlinstown Wood / Loughlinstown

Commons, visible just north of Commons Road. A millrace and corn mill once existed in this area, and flooding in not uncommon here. The river then travels under Shanganagh Bridge. The final reaches of the river curve northeast (above right) and it passes through the railway embankment (as seen on the cover) and runs north of a local authority water purification

plant. The Shanganagh then winds across the gravelly beach to the shingly Killiney Bay (shown left), north of the Shanganagh Cliffs area.

Note: In mapping up to the first edition of the Ordnance Survey, including Taylor's map of 1816, the final stages of the Shanganagh run north along the coast, instead of directly to the sea, and take in the waters of the Kill o'the Grange Stream before finally turning to the sea just south of the Killiney Bay Martello Tower. How the two watercourses' mouths became separated is not now clear.

There was historically a short stream rising just southeast of the southern Shanganagh Castle and coming to the sea east of the two Shanganagh (rail) Junction locations. It ran, and some water still appears to flow to some degree, southeast and west within castle grounds (now in public hands, with a major development underway). It goes on south, through what is now Shanganagh Park, receiving a small flow from muddy ground at the southern edge of the park. It runs on southeast, receiving a tributary along the eastern edge of the western Shanganagh Cemetery lands. This tributary comes from a ditch running from opposite the former St James' Parsonage (now Ask(e)field), and a stone works, to St James' Church, and a line from its mid-point, along the southern edge of cemetery lands. It crosses the modern railway near a small bridge, and goes east and north. Its final short run east used to cross two railway lines, the original final stretch of the Harcourt Street railway and the coastal railway, the first line of the Dublin to Wexford railway, and it then came to the sea. The coastal line, of which little trace remains, was rendered unsafe due to coastal eroson, and in 1915 it was replaced by the modern line, designated the Shanganagh Deviation, from Ballybrack to near Bray. Remains of the former line are now a striking feature in the surf from Shanganagh to near Bray.

- **Crinken Stream** (also **Wood Brook**, and possibly **Allies River**), from Rathmichael and

western Shankill In most modern mapping it is shown forming from two flows near Shankill Castle, and running south and southeast, to the sea a little north of the town of Bray. Historically the Crinken began with a stream from south of Puck's Castle, just below some ancient earthworks, which flowed east, taking in a branch from the southwest near Shankill Castle, then a joining of two additional southern lines. It then went northeast (see left), and, just west of the house called Lordello, took in a modest flow from a pond east of Shankill Castle. This confluence is where modern mapping begins; of the old feeder lines, at least two, including the longest ("the Puck's Castle branch"), still exist to some degree.

From the merger south of Lordello Road, the Crinken flows southeast, crossing Ferndale Glen and running south of Ballybride Road, then east to go under the M11 Dublin-Wexford road parallel to and just south of Crinken Lane, which overflies the M11. It goes on southeast, under Allies River Road, where it receives a tributary at the rear of the former parsonage of St James. This streamlet comes from between the Old Conna Hill estate (nowadays a golf club) and the former *Toole's Moat* earthwork. The Crinken continues through the lands of the house latterly named The Aske and turns sharply east to pass a school and cross Dublin Road north of the motorway exit.

It then enters Woodbrook estate lands (see right). It runs along the boundary of the inner estate, parallel to the road to an auction house and one of Ireland's older golf courses, passing Corke Lodge with its ornamental gardens. Near where there was a cricket pavilion and an opera hall, and a special railway stop, it forms a pond. Exiting this to the south, it goes into a short culvert as it approaches the road

called The Fairways. Emerging from under that road, in a valley between

roads called The Drive and The Grove, and running east of Corke Abbey, the Crinken runs roughly southeast and east, mostly in the open, in a developing public park (shown to left).

The stream's line was diverted southwards in the 20th century to merge with (taking over the lower stages, where the upper were lost) a small stream coming from Old Connaught Avenue via Corke Abbey (and that used to take in a flow from the Abbey Well), sometimes *Old Connaught Stream*. Corke Abbey, the site of a monastic development, was in the 20th century known for the Solas Teoranta light bulb factory.

The two streams used to come to the sea in parallel but with Crinken diverted into the southern stream's course for its last few dozen metres, the combined flow goes to and under the railway and then via soft ground to reach, untidily, the narrow stony strand between Shanganagh Cliffs and Bray, and so to come to the sea just north of the River Dargle's mouth.

The stream's catchment, earlier altered by works on Rathmichael Road, was further modified during the construction of the final M50 section, with some diversion of water to the Bride's Glen Stream part of the Shanganagh River system set up so as not to overload drainage in the southern part of modern Shankill village.

- **Glenmunder Stream** (also known as **County Brook** as it marks part of the border with Wicklow, or **Ballyman Stream** from the eponymous valley, or **Fassaroe Stream**) flows south from Ballybetagh (east of Glencullen in an area mostly draining to the Glencullen River, discussed below), then east, eventually passing between a golf hotel and a skiing facility to turn to flow south through The Scalp. It goes south & southeast from Phrompstown through Ballyman, passes Kevin's Well, near which it may receive a small northern stream, and after Fassaroe it runs south of Vallombrosa (near which lay another well) and comes close to the Dargle River in northern Wicklow, on the outskirts of Bray. It flows parallel to the river (see right) and Upper Dargle Road for a short way, and then merges in near Coburg and the former sites of Charles' and Broderick's Wells.

- **Glencullen River** *(Abhainn Ghleann Cuilinn)* flows from midway along the southern Dublin valley called Glendoo to the northwest and Glencullen to the southeast. It takes in many tributaries, the larger including, from the southwest, **Middle Brook** (the central one of three streams) on the slopes of Glendoo, and **Glanduff** from Knocknagun, nearer to Glencullen. There is also a flow from Ballyedmonduff, past Glencullen, coming from the northeast. The Glencullen then runs to the south of Glencullen village, and meets **Brockey Stream** (*Glasnabrockey*), along which the county boundary runs before following the main river for a short way and then heading north along the line of another tributary, from Ballybetagh, leaving the Glencullen in Wicklow.

The river flows through Knocksink Glen and Woods as the **Cookstown River** (*An Chlóideach*) (left below). It passes Enniskerry (right below) before merging into the Dargle (which has run by Powerscourt Demesne and picked up a stream) near the Dublin-Wexford road.

The Glencullen has a healthy population of brown trout and is also a spawning locale for salmon and sea trout.

= = = Gazetteer ends = = =

The author

Born by the Dodder, and growing up between the Santry River and Fox Stream in Raheny, with the Naniken close by in St Anne's Park, Joe Doyle works in a high-tech business totally unrelated to rivers. An attempt to figure out what the Fox Stream was, at the age of 11, with the aid of a fascinating local history file at Raheny Library, ultimately led to this book. A graduate of Dublin City University and NUI Maynooth, Mr Doyle has worked in several, and visited more than 65, countries. Married with two daughters, he has plans for a book on the waters of another great city (more than 220 rivers and streams identified), but also for a future edition of this work with refreshed photography, and more mapping.

The Rivers and Streams of Dublin (City of Dublin, Fingal, South Dublin, Dún Laoghaire-Rathdown)

End Notes

Publication history and plans

This book descends from a leaflet I wrote during 1999's burst of Millennium-related activity that, as a first on the topic at county level, attracted some interest. Years later, fresh research allowed better coverage, including naming more watercourses, and noting more of the "now nameless." Each edition (the way ISBNs work, changes in page count create new editions) allows input from site visits, and more photos, and newly available maps, tools, and sources. The historical images are from prints and postcards I own, while the modern are my own amateur shots from recent years. This is a non-profit project, largely written while living abroad, and progress on images and mapping takes time (to map properly, at a scale where the smaller watercourses would be clear, is hard). Initial material for this project was gathered over a period in 2008, from libraries and historical materials held personally. Over time many points were checked manually - most watercourses were visited at one or more locations, some quite obscure - and some re-evaluated. Questions remain, however - any new information is very welcome (by e-mail to rer@raheny.com, and I hope via a website for the book at some point, or a Facebook or similar page). For those "tracking down" waters, there may sometimes be disappointment as some in urban parts no longer look very, as you might say, stream-like, and many watercourses have been partly canalised, or integrated, in rural areas, with drainage channels, though winding courses can help to distinguish them. Some streams were already "in trouble" decades ago, such as several Dodder tributaries between Glenasmole and Templeogue, left short of water due to changes in land usage, and the courses of some exist now only as "wet ditches", with their main flows diverted. Old maps and tracing paper, and official drainage information, can assist, though it is not always easy to identify stream lines on old maps absolutely where they were used to set boundaries and the markings for these confuse the reading.

Sources

A modest bibliography is included, and a simple index of watercourse names. Sources include the various books mentioned, especially Sweeney's "Rivers of Dublin," and my leaflet and surviving notes, maps (especially Rocque, Duncan, Taylor, Ordnance Survey, and drainage maps, as well as, for later editions, Google Maps and similar tools), local authority material (incl. planning records, County / City and Local Area Plans, meeting records, etc.), flood reports, newspaper archives (chiefly as secondary sources), Oireachtas debates, Statutory Instruments, Environmental Protection Agency materials, and local history publications (though watercourses, except the Poddle, Liffey and City Watercourse, seem not to be popular topics). For example, the little Kill o'the Grange system was traced with a mix of 19th and 20th century maps, site visits (most of the main course directly south from Kill o'the Grange is in the open) and archives (e.g. Dáil mention of culverting of the Deansgrange Stream). There remain uncertainties, even contradictions, such as on source points (mountain streams often having more than one), and while the bulk of naming is clear, there are cases with lasting unclarity, partly due to the fact that when people moved around less, many watercourses had different names in different parts. There are also simply errors - for example, where several tributaries exist, a report may miss one, "advancing" later names from a list - or when Kill o'the Grange Stream is listed as a tributary of the Shanganagh, something which anyone who has walked Killiney Strand knows to be incorrect, at least today - but which *may* have been so once (similar queries with regard to other waters were resolved). For later editions, "Hidden Streams" by an established local historian of Dún Laoghaire-Rathdown has been of great interest, along with several Paddy Healy publications brought out by South Dublin Libraries in the 2000s, deCourcy's scholarly "The Liffey in Dublin," "The Book of the Liffey from Source to Sea," by Healy, Moriarty and O'Flaherty, and Moriarty's excellent companion to any walk by the river, "Down the Dodder" (a similar companion to the Tolka would be very welcome).

The nameless and "the maybes"...

Some of the unnamed watercourses here may well have had names, or even now have some very local naming; many could safely be named for a location through which they flow, e.g. Abbotstown Stream. In the *Rivers of Dublin*, Sweeney assigns "working titles" to branches of the Swan Water, and some other streams, and similarly, for example, the Ward River could be listed with tributaries such as Lubbers Wood Stream, the Tolka a Castaheany or Mulhuddart Stream, the Little Dargle the Grumley's Well Stream, and the Liffey one or more Palmerstown Streams. There are also a few "unconfirmed" waters, including two "conjectural" items in Sweeney's book ("Gallows Stream" and "Oxmanstown Stream"), as well as small flows marked on some maps but not otherwise traceable in records, or visible - such as short water-courses on historical maps of Clontarf (as they are long buried, it is not clear whether they were natural), the "Portland Watercourse" in the inner city (possibly a stretch of the Royal Canal), and Fingal's "Middle Stream." Such are not necessarily in the main text but if more information is found, can be detailed in a future edition. There are also "disrupted" watercourses, such as the Marino Stream, their status unclear.

Canals, drainage and water supply, bridges and mills…
There are related topics on which it would have been interesting to write, such as the canals, or the water supply or drainage systems - but each could fill a whole book (see "Our Good Health" for great detail on water supply and drainage, for example). The canals feature on occasion, but were usually kept detached from natural watercourses, and the same logic is followed here (including for canal "streams" such as the "Cappagh Overflow"). Water supply comes up at times, notably related to the Dodder, Poddle and Camac in earlier times, and drainage appears on several occasions, most notably with regard to the GDDS (Greater Dublin Drainage System), specifically the GTS (Grand Canal Tunnel Sewer), now capturing the main flow of several old waters. Another topic worth elaborating in a longer book would be the bridges over the rivers and streams - tracing just one watercourse can yield a whole range of interesting names, with linked history. And then there are curiosities, such as the stream and pool detailed in an academic paper, feeding both the Owendoher (north-bound) and the Glencullen River (south-bound). And worthy of their own volume, but mentioned here only in passing, are the mills of County Dublin, mostly gone now, and its holy wells (Fingal has a fine book by Petra Skyvova on the topic).

Other cities
Some readers have asked about possibilities for other areas of Ireland and even the U.K. I did study the rivers of several other parts of Ireland, and of London (which has a fine set of hidden and half-forgotten waters, though two good books and a series of best-selling fantastic novels have made people more aware of some) and of Leeds and Sheffield. I have no plans for even leaflets for Great Britain for now, but as even the basics of Cork City beyond the Lee, and Belfast beyond the Lagan, are not well known, include here brief notes on these two for a start (there are *very* few city watercourses in Galway or Limerick beyond the powerful main rivers, but as for counties, there are so many…).

Rivers of the city of Cork
While the Lee dominates, and its maze of channels and islands formed the historic urban core, there have always been other waters, without which Cork would have been very different (in particular, it might not have had such a position in the island's brewing, distilling and tanning industries). One even runs almost parallel to St Patrick's Hill. As it approaches the city's centre, the Lee is today divided into northern and southern channels. The southern channel receives a combined tributary line just as it heads into the centre, at the western end of the UCC campus. This line is historically sometimes mapped as the Maglin River, but is today better known as the Curraheen River; these are the names of two of its three main contributors, the third being the Glasheen. The Maglin comes from furthest west, from beyond Ballincollig Castle, and runs roughly east, meeting the Curraheen (or Curragheen), which passes the area of the same name but comes from branches starting well to the south, including towards Ballinhassig. These two join together by a junction on the N40 and form a substantial flow, which receives the Twopot River north of Curraheen Estate, by Murphy's Farm and the Cork Institute of Technology campus. It runs on north, goes under Model Farm Road, turns east and tends towards Carrigohane Road, which it then roughly parallels for a stretch. Just downstream of the N71, the Curraheen takes in the Glasheen River as they both approach the Lee. The Glasheen rises to the south and flows roughly but fairly directly north to near the Bandon Road roundabout on the South Ring Road, then east, parallel to that road, to near Deanrock. It then goes north, northwest, and north again from Dennehy's Cross to join the Curraheen just downstream of Victoria Bridge. Together, these modest rivers drain much of Cork city's western and southwestern hinterland.

Deep in central Cork, what is historically mapped as the Kiln River, and often today as the Bride River, flows in to the Lee just upstream of the Christy Ring Bridge. This forms in stages, with rural origins, with the Bride River from the west, and Glennamought (or Kilnap) River from the east, meeting at Kilnap (each in turn has multiple preceding branches), proceeding as one through the Blackpool Valley, and then in turn receiving the Glen River (from the former Goulding's Glen to the east). The line of its last stages runs parallel to the North City Link Road, where a small part may still be traced, crosses Devonshire Street, and reaches the Lee at Camden Quay. The lower reaches and connected watercourse lines once served the Watercourse Distillery and Lady's Well Brewery, and part of these reaches, and their immediate area, were once known as "*Poweraddy*."

Running across the southern stretches of the city, taking tributaries from north and south, the Tramore River rises in Gortagoulane, a little north of Cork Airport, flows roughly north, passes the site of a former railway line, and comes to the Fernwood development. It goes on northeast a bit, then north again to the Togher Road just south of the South Ring Road. It then turns to flow east parallel to the South Ring Road, here following the line of another former railway, and crosses to run on the road's north side before the Kinsale Roundabout. The Tramore receives a small tributary from the Frankfield direction, and soon after another from south of Marian Grange, passing Vernon Mount. It then winds on east and northeast, crossing the ring road twice, running just northwest of Douglas Shopping Centre and receiving the Trabeg River. The latter has come north from branches near the site of Castle Treasure, from south of Bramble Hill, by Ballinvuskig, and from Graigue a little to the west. The Trabeg has then gathered and run a little west and north again, passing Cork's Donnybrook and Ballybrack, and at least one former mill site. The Tramore or Douglas River then goes a little northeast to its estuary in the Douglas Channel, eventually opening out into the early stages of the Lee estuary at the northern end of Monastery Road, west of Hop Island, where a last tributary joins from Rochestown. This latter comes from branches in Ballyorban, Oldcourt and Rathanker beyond southeastern Rochestown.

Just outside the city proper, the Glashaboy River comes to the beginning of the Lee estuary, running south of Glanmire, and touching the edge of urban development; it has come south and southeast, with one major tributary, the Butterstream River, joining north of Glanmire. It forms from many branches in the Glashaboy and Glynn areas, and earlier tributaries include the busy Cloghnagashee River which flows in from Carrignavar and which takes in the Glantaunkeagh River, and the Black Brook (joining north of Ballinvriskig, just above Upper Glanmire Bridge).

Rivers of Belfast

If asked the primary name of "the Belfast River" most from outside Belfast would probably name the Lagan. And the Lagan is by far the main watercourse of the Belfast area. But the city's name in Irish, *Béal Feirste*, points the way to the correct answer - the now-hidden River Farset (also known as the Town River and the High Street River), from branches in the Squire's Hill area. The city is not named for the river, rather both river and city are named from the Irish for "sandbar".

The Farset line is reflected in that of Belfast's High Street, up which boats could move until the late 18th century (the river was only fully covered in the mid-19th century). It rises on Squire's Hill. In turn, the Farset is just one of several waters falling to the Lagan or its estuary from the hills on the western and eastern approaches to the city, which in turn have dozens of tributaries, many quite tiny and many also long ago culverted or diverted (some not otherwise discussed include the Altona and Glentoran Streams, the Carr's Glen River, the Three Mile Water, and Tillysburn).

Just to the south of the Farset, the larger Blackstaff River or Owenvarra(gh) flows in (sometimes the "Blackie" locally); the city developed around the meetings of Farset, Blackstaff and Lagan. The city centre stages were culverted in an 1881 plan, but much of the Blackstaff system can still be found overground, and it has caused flooding even in recent times. The Blackstaff comes from west and south of the city, from the lowlands overlooked by Black Hill (branches include the Falls Road Stream, Ladybrook River and Ballymurphy Stream), meeting the Clowney Water at the Broadway Roundabout of the Westlink, a meeting much reworked in the last decade, and which includes a modern relief diversion to the Lagan south of the main outfall. The Blackstaff continues with the Westlink for a time, then turns towards the Lagan, having taken in the Pound Burn (linking to the Farset). It crosses Durham Street and Great Victoria Street, and eventually passes Donegall Square West, goes along Donegall Square North and angles from Chichester Street to cross Victoria Street and Oxford Street to its large culvert exit. The Clowney Water is the end of the Forth River line, with inflows from Divis, Glencairn and Crow Glen forming the Forth, then boosted by the Ballygomartin River nearer the city, and other lines.

Parts of eastern Belfast drain to the Connswater, partly parallel to the Lagan, in turn forming from the Knock and Loop Rivers, the former from near Stormont (with tributaries including Knock Burn and Gilnahirk Stream), the latter from Cregagh and Lisnabreeny via Castlereagh, taking in the Glenbrook River and Merok Burn. South of all these the Glencregagh runs west to the Lagan.

A little further north, from the west, the Milewater joins, from springs at Squire's Hill at Ballysillan, and elsewhere near Cave Hill, and with its historic mouth around where Yorkgate Station stands, and further north again several short streams run east to the estuary, while from the east, around Hollywood, there historically ran streams including The Glen and The Stinking Burn.

South of the city core, there are streams in the Belvoir Park area, and the Purdysburn, having taken in the Carryduff, flowing in at Minnowburn, and also the Deriaghy, and the Colin River.

Note: At some point I hope to draw on the excellent *The Rivers of Belfast* and other sources, after a visit for some photography, for a longer piece on the city and surrounds.

Partial bibliography

Dublin: Ball, Francis Elrington, *A History of the County Dublin* (6 volumes, 1902-1920)
Dublin: Byrne & Graham, *From Generation to Generation - Clondalkin Village, Parish and Neighbourhood*, 1989
Dublin: Cawley et al., *A Selection of Extreme Flood Events - The Irish Experience*
Dublin: Centre for Water Resources Research (UCD) *Deliverable 10.3: Assessment of Factors Affecting Flood Forecasting Accuracy and Reliability* (Draft), 2004
  (Bruen with Gebre, for City Council) *An investigation of the Flood Studies Report ungauged catchment method for Mid-Eastern Ireland & Dublin*, 2005
Clane, Co. Kildare: Clane Local History Group, *Hidden Gems and Forgotten People*
Dublin: Coillte - *Forest Management Plans*
Dublin: Four Courts Press / Dublin City Council, Corcoran, Michael, *Our Good Health - a history of Dublin's water and drainage*, 2006
Dublin: Hodges & Smith, D'Alton, John, *The History of County Dublin*, 1838
Dublin: Gill & Macmillan, de Courcy, J.W. - *The Liffey in Dublin*, 1996
Dublin: Cosgrove, Dillon, *North Dublin*, 1909
Dublin Airport, Co. Dublin: Dublin Airport Authority, *Your Airport (Issue 2)*, 2006
Dublin: Dublin City Council (historically Dublin Corporation) - various, e.g.
  City Engineer's Dept. - Sewers Section, *Annotated OSI (1958-1959) map - General Plan of Storm Section*
  City Engineer's Dept. - *Map "Courses of Poddle River and City Watercourse"*
  Sutton to Sandycove Promenade and Cycleway (various documents)
  (Collins and McEntee) *A Constructed Wetland for the Removal of Urban Pollution in the Finglaswood Stream, Tolka Valley Park, Dublin*, 2009
  (Conway, Adrian P.) *The North Fringe Project - Delivery of Sewer and Water Infrastructure to North Dublin*, 2006
  (Frazer, William O.) *Newmarket and Weaver's Square* (MosArt) *Bushy Park: Landscape Masterplan and Management & Development Plan*
  City Development Plans (inc. *Environmental reports*)
  Local Area Plans
Dublin: Dublin Co. Council - various, e.g. flood reports
Dublin: Dublin Drainage Consultancy, *Greater Dublin Strategic Drainage Study*, 2003
Dublin: Dublin Drainage Consultancy / M.C. O'Sullivan & Co. Ltd., *River Tolka Flooding Study* (extension to the Greater Dublin Strategic Drainage Study), 2002
Dublin: Dublin Regional Authority - various, e.g.
  Sutton to Sandycove Promenade and Cycleway
Dún Laoghaire, Co. Dublin: Dún Laoghaire-Rathdown County Council - various, e.g.
  Water and Drainage Division - *Hand-annotated map of Dun-Laoghaire-Rathdown*
  Treasuring Our Wildlife - Dún Laoghaire-Rathdown Biodiversity Plan 2009-2013
  Implementation Report for the Phosphorus Regulations
  Dublin Road Improvement Scheme, Bray, Environmental Impact Statement, 2008
  Water Services Investment Programme - Assessment of Needs (draft), 2003
  County Development Plans (especially *Environmental reports*)
  Local Area Plans (Stepaside and Ballyogan, Kilternan, Glencullen, etc.)

Dublin: Duncan, William, *Map of the County of Dublin*, 1821
Dublin: Eastern Regional Fisheries Board - various, e.g. *Annual Reports*
Dublin: Eastern River Basin District - various, e.g. *Draft Programmes of Measures* (by Water Management Unit)
Clonegal, Co. Wexford: EastWest Mapping, *The Dublin & North Wicklow Mountains - A Detailed Map*, 2009
Dublin: Environment, Dept. of - various, e.g. *Water Services Investment Programme*, excavations.ie
Johnstown Castle Estate, Co. Wexford: Environmental Protection Agency - various, e.g.
  Interim Report on the Biological Survey of River Quality (various years and volumes)
  Wastewater Discharge Licence, Recommended Decision - Shanganagh
Dublin: ERM for the OPW and local authorities, *Towards a Liffey Valley Park - Strategy Document*
Dublin: ESB International for Eirgrid, *Interconnector Project - Ireland Land Environmental Report*, 2008
Swords, Co. Dublin: Fingal County Council - various documents, e.g.
  Abbotstown Sports Campus Study, 2007
  Architectural Conservation Area documents
  County Development Plans (especially *Environmental reports*)
  Local Area Plans (Balbriggan North, Diswellstown, Donabate, Dublin Airport, Mooretown-Oldtown, Portmarnock, etc.) and related assessments, etc.
Dublin: Geological Survey of Ireland - various *Geological Site Reports*
Glenageary, Co. Dublin: Glenageary, Catholic Parish of, *Parish History* (online)
Dublin: Grangegorman Development Agency, *Grangegorman: An Urban Quarter with an Open Future - Utilities and Infrastructure*
Dublin: Halcrow Barry for Fingal County Council and the OPW, *Fingal East Meath Flood Risk Assessment and Management Study, Environmental Scoping Report*, 2009
Dublin: Handcock, William Domville, *The History and Antiquities of Tallaght in the County of Dublin*, 1899 (2nd ed.) [Tallaght refers to the civil parish, not the then village]
Tallaght, Co. Dublin: South Dublin County Council, Healy, Patrick: *All Roads Lead to Tallaght*, 2004; *Rathfarnham Roads*, 2005; *Glenasmole Roads*, 2006
Dublin: Wolfhound Press, Healy, E., Moriarty, C., O'Flaherty, G., *The Book of the Liffey from source to the sea*
Dublin: The Institution of Engineers of Ireland, McDaid et al., *South Eastern Motorway - The Road Delivered*
Dublin: The Irish Naturalist - various, e.g.
  Vol. 1, no. 5, Grenville A. Cole, *County Dublin, Past and Present* (concluded)
  Vol. 11, no. 4, Wright, W.B., *The Glacial Origin of Glendoo*
Dublin: The Irish Independent - various articles
Dublin: The Irish Times - various articles and letters
Dublin: Joyce, Weston St John, *The Neighbourhood of Dublin* (various editions, up to 1920)
Naas, Co. Kildare: Kildare County Council - various documents, e.g.
  County Development Plans (especially *Environmental reports*)

The Rivers and Streams of Dublin (City of Dublin, Fingal, South Dublin, Dún Laoghaire-Rathdown)

Local Area Plans (Donaghcumper Lands - Celbridge Town Centre, etc.)

Dublin: National Library of Ireland, Kissane, Noel, *Historic Dublin Maps*

London: S. Lewis & Co., Lewis, Samuel, *A topographical dictionary of Ireland…* (2 vol's, with map book), 1837

Dublin: Currach Press, Mac Aonghusa, Brian, *Hidden Streams: A New History of Dún Laoghaire-Rathdown*, 2007

Dún Laoghaire, Co. Dublin: Dún Laoghaire - Dublin's Riviera Ltd., Merrigan, Michael, *Proposals for Sustainable Development*, 2001

Monkstown, Co. Dublin: Monkstown, Catholic Parish of, *Parish History* (online)

Dublin: Wolfhound Press, Moriarty, Christopher, *Down the Dodder*, 1991

Dublin: National Parks and Wildlife Service - various SPA materials

Dublin: National Roads Authority - EIS materials

Cork: O'Callaghan Moran & Associates, *Annual Environmental Report for Greenstar Materials Recovery Ltd. Fassaroe Depot (Licence No. 53-2) 2003-2004*, 2004

Dublin: Office of Public Works - various, e.g.
  Flood Hazard Mapping collections and minutes
  The Phoenix Park Conservation Management Plan - Consultation Draft, 2009
  Visitor material at Castletown House, Maynooth Castle, Phoenix Park Visitor Centre, etc.

Dublin: Oireachtas, Houses of the - various, including record of debates

Dublin: Old Dublin Society, Dublin Historical Record - various, e.g.
  Vol. 2, no. 2 - Hegarty, James, *The Dodder Valley*
  Vol. 3, no. 1 - O'Brennan, Lily, *Little Rivers of Dublin*
  Vol. 11, no. 1 - Jackson, V., *The Glib Water and Colman's Brook*
  Vol. 12, no. 4 - Holden, P.J., *Local Taxation in Dublin, 1812*
  Vol. 15, no. 2 - Jackson, V., *The Inception of the Dodder Water Supply*
  Vol. 29, no. 4 - Dawson, T., *The Road to Howth*
  Vol. 31, no. 4 - Flood, Donal T., *Dublin Bay in the 18th Century*
  Vol. 47, no. 1 - Behan, A.P.: *Up Harcourt Street from the Green*
  Vol. 58, no. 1 - Clare, Liam: *The Kill and the Grange of Clonkeen: Two Early Settlements in South Co. Dublin*

Dublin: Parliament of Ireland - various, e.g. Acts re: the Poddle River (17 & 18 Car.2 [I], 36 G.3 [I])

London: Parliament of the United Kingdom - various, e.g. An Act to Amend the Acts Relating to the River Poddle in the County and City of Dublin (Cap. LVIII, 3 & 4 Vic.)

Dublin: Scoil Treasa Publications, Ó'Néill, Séamus, *Firhouse*

Dublin: Ordnance Survey Ireland - maps (from 1837 to present), including their excellent online "viewer" tool

Raheny, Dublin: Raheny Online Working Group (Author: J. Doyle), *Rivers and Streams of County Dublin*, 1999

Dublin: Railway Procurement Authority - various re. Luas, Metro North, Metro West

Dublin: Rocque, John, *An actual survey of the county of Dublin on the same scale as those of Middlesex, Oxford, Barks and Buckinghamshire* (4 sheets), 1760

Dublin: Royal Irish Academy - various, e.g.
  Proceedings of the Royal Irish Academy. Section C, Vol. 62 - O'Conbhuí, C., *The Lands of St Mary's Abbey, Dublin*

Dublin, Ireland, Royal Society of Antiquaries of Ireland, Journal of the RSAI - various, e.g.
  (JKSEIAS, New Series, vol. 2, no. 2, *Proceedings and Papers*)
  5th series, vol. 10, no. 4 - Ball, Francis Elrington, *The Antiquities from Blackrock to Dublin*
  5th series, vol. 32, no. 2 - O'Reilly, P.J., *Tobernea Holy Well, Blackrock, County Dublin*
  7th series, vol. 12, no. 3 - Ua Broin, Liam, *Traditions of Drimnagh, Co. Dublin, and Its Neighbourhood*
  Vol. 78, no. 1 - Ronan, M.V., *The Poddle River*
  Vol. 83, no. 1 - Lucas, A.T., *The Horizontal Mill in Ireland*
  Vol. 87, no. 1 - Ua Broin, Liam, *The Mountain Commons of Saggart*

Dublin: S2S, *Online materials re. Sutton2Sandycove proposed promenade*

Dublin: SEMPA, *Report from Local Planning Group (LPG) 2 - Recreation and Tourism Plan for North Dublin Bay*, 2001

Swords, Co. Dublin: Fingal County Libraries, Skyvova, Petra, *Fingallian Holy Wells*, 2005

Tallaght, Co. Dublin: South Dublin County Council - various documents, e.g.
  (Murray, Sean) Urban River Management, 2000
  County Development Plans (especially *Environmental reports*)
  Local Area Plans
  Connecting to the Past: Mapping South Dublin County in Time (a great online tool, accessing multiple historic maps)

London: John Sudbury & George Humble, Speed, John, (Inset map of Dublin from) *The Countie of Leinster with the Citie Dublin described*, 1610

Dublin: Stillorgan Chamber of Commerce, *The History of Stillorgan* (online)

Dublin: Dublin Corporation, Sweeney, Clair L., *The Rivers of Dublin*, 1991

Dublin: Taylor, John, *Taylor's Map of the Environs of Dublin extending 10 to 14 miles from the castle, by actual survey, on a scale of 2 inches to one mile*, 1816

Dublin: Project Group of the Templeogue Ladies Club, *The Story of Templeogue*, 1992

Terenure, Dublin: Terenure College, *Lake Wildlife Walk*

Wikipedia, various articles and discussion pages

And in addition:
  For all related local authorities and An Bord Pleanála, various planning applications and other related material
  For all related local authorities, Minutes of Meetings (especially area committees) and other materials
  Some Land Registry and Registry of Deeds materials

Biographical notes

Clair L. Sweeney, draughtsman, author and artist (1923 – 1997)

Clair Louis Sweeney was born in 1923, and grew up in Palmerstown (Palmerston), in the Liffey Valley, by the Strawberry Beds. At that time, Palmerstown and the 'Beds were distinct from a much smaller Dublin. He worked for Dublin Corporation over decades, notably for the Drainage Department, and with his draughtsman's skills, produced many drawings of the city's watercourses and drainage lines. He also developed expertise in searching archives and was a keen student of old maps. Consulted when surface installations were to be made, including the planting of trees, and underground workings dug, from Kilmainham to Ballymun, Palmerstown to Raheny, he was, for example, involved in the location of the site of St Winifred's Well in Temple Bar. In 1988 he was the recipient of the Lord Mayor's Award for his services to the city. He retired from the Corporation in 1990.

Clair Sweeney wrote the first major book on the rivers and streams of Dublin, specifically on those passing through the city, in a project inspired by Dublin's Millennium. Published in 1991 in large-format paperback, with a small run of hardbacks, "The Rivers of Dublin" is a classic, with fine maps and illustrations. A revised edition of his book, with some updating by two fellow City Council staff, was published in November 2017.

In later life, he took up oil painting, and sold his landscapes on city park railings. He was an active member of Palmerstown Parish (R.C.), helping to restore statues and Stations of the Cross, and to build the grotto, at St Philomena's Church.

Mr Sweeney was married to Eileen (-2017); they had three children, Delma, Pascal and Colm. He died at Hampstead Hospital on June 15, 1997. His remains were moved to St Philomena's Church on Bloomsday, and after his funeral on June 17, he was buried in Palmerstown Cemetery. He was remembered with an Irish Times obituary on the 25th July, 1997.

Christopher Moriarty, ichthyologist / naturalist, historian and author (1936 – 2024)

Born March 14th, 1936, Christopher Moriarty grew up in a Rathfarnham then still remote from Dublin city, and attended St Columba's College, where he spearheaded the growth of the school's Natural History Society. He secured a bachelor's degree in zoology at Trinity College Dublin, and joined the public service, becoming a national, and later international, authority on eels. He was Ireland's liaison to a commission of the Food and Agriculture Organisation (UN) for 40 years, chairing it for three years and leading its working group on eels for around a decade. He took a two-year assignment at Ibadan University in Nigeria, to set up a fisheries course.

He studied for a Masters at Trinity, investigating trout and perch at Poulaphouca Reservoir, and in 1972 completed a TCD Ph.D., on eels (later the basis for a book). Long active in the Dublin Field Naturalists Club, he was its President for a time in the 1970s. A long-term member of the Royal Dublin Society (RDS), he spent some years on its scientific committee.

He was married to Sue Goldie (-2012), who studied and published on Florence Nightingale; they had two sons, Patrick and Ruairi. In later life, his partner was Mary Pyle of Palmerston. He was a long-time resident of an apartment at the historic house at Woodtown Park, on a history of which he was working at the time of his death, on 13th January, 2024. An active member of the Society of Friends (Quakers) since the 1970s, he was Clerk of the Dublin Meeting for three years, and of the society's national historical committee for many years, as well as curator of its Irish historical library. He was buried, after a Meeting for Worship on January 18th, 2024, at the Quaker Cemetery in Blackrock, Dublin.

Dr Moriarty wrote for both the Evening Press and the Irish Times. He was the author of a range of accessible and popular books on nature and travel, beginning with "A Guide to Irish Birds" (1967), and the compact "A Natural History of Ireland" (1971), followed by 1978's "Eels: A Natural and Unnatural History". He also wrote multiple volumes of the "Irish Environment" series, widely used in schools and libraries for many years, as well as short books on cycling and driving routes, and a volume on Dublin parks and waterways. Specifically on the watercourses of Dublin, he was lead author for "The Book of the Liffey", and sole author of "Down the Dodder", often cited as the best book on the main line of that river (and latterly made available as an e-book), and of 2018's well-illustrated "The River Liffey History and Heritage". He was the subject of an Irish Times obituary on 10th February, 2024.

The Rivers and Streams of Dublin (City of Dublin, Fingal, South Dublin, Dún Laoghaire-Rathdown)

Index of watercourse names within County Dublin
- alphabetic; includes main, alternative, and not fully confirmed; page number is that of primary appearance of name
- does not include tributaries outside Co. Dublin (but see next page for Liffey tributaries in Counties Wicklow and Kildare), nor the supplementary articles on Cork City and Belfast

The natural watercourses

Name	Page	Name	Page	Name	Page
Abbey Stream (Rathfarnham)	74	Coombe Stream	61	Griffin River	49
Ailesbury Stream	80	Corbally Slade River	53	Hampstead Stream	36
Allies River	87	Corduff River (Stream)	8	Hazelbrook Stream	18
Allison's Brook	65	Cot Brook	66	Hollybrook Stream	29
Anna Livia	38	Coulcour Brook	22	Hurley River	3
Ardla Stream	5	County Brook	88	Inch Stream	4
Ath Collop	53	Creosote Stream, The	51	Jobstown Stream	68
Badger(s) Glen Stream	73	Crinken Stream	87	Jone's Stream	6
Baker's Well Stream	50	Crooked River, The	52	Kamoke River	52
Balcunnin Stream	6	Cruagh Stream	71	Kealy's Stream	17
Balleally Stream	7	Crumocke River	52	Kellystown Stream	50
Ballinascorney Stream	67	Cuckoo Stream	18	Kenure Stream	6
Ballinclea Stream	82	Daunagh Water	23	Khyber Stream	52
Ballough Stream	8	Daws River	7	Kilbarrack Stream	23
Ballyboughal (Ballyboghil) River	7	Dean's Grange Stream	83	Kilbride River	39
Ballycullen Stream	70	Deansgrange Stream	83	Kildonan Stream	33
Ballyhoy River (Stream)	26	Delvin, River	2	Kilgobbin Stream	85
Ballymaice Stream	67	Delvyn (River)	3	Kill o'the Grange Stream	83
Ballyman Stream	88	Diswellstown Stream	50	Killakee Stream	71
Ballyogan Stream	85	Dodder, River	65	Killeen Water	53
Balsaggart Stream	22	Donabate River	9	Killinarden Stream	68
Bartramstown River	2	Drimnagh Castle Stream	54	Killininny Stream	68
Belinstown Stream	9	Dun Water	10	Kilmacud Stream	81
Blackbanks Stream	24	Dundrum River	74	Kilmahuddrick Stream	49
Blackditch Stream	54	Dundrum Slang	74	Knockmaroon Stream	50
Bloody Stream, The	21	Dunshaughlin Stream	10	Lane Stream	5
Bluebell Stream	54	Elm Park Stream	78	Leopardstown Stream	81
Boddeen Stream	21	Elvene (River)	3	Leper's Stream	81
Boggeen Stream	21	Elvin Water	3	Liffey, River	38
Boherboy Stream	53	Elvin, River	3	Lisheens River	39
Bracken River	4	Esker River	49	Lissenhall Stream	9
Bradoge River	56	Fassaroe Stream	88	Little Dargle River	73
Bremore River	4	Ferny Glinn	53	Loughlinstown River (north or	
Brewery Stream	81	Fettercairn Stream	53	Carrickmines branch)	85
Bride's Glen Stream	86	Finglas River	33	Loughlinstown River (south or	
Bride's Stream	6	Finglaswood Stream	33	main branch)	86
Brittas River	39	Finisk Stream	55	Lusk River	7
Broad Meadow Water		Forrest Little Stream	17	Mabestown Stream	12
(Broadmeadow River)	10	Fox Stream	24	Magazine Stream	52
Brockey Stream	89	Fox's Lane Watercourse	24	Maine River	18
Brook Stream (Rush)	6	Foxrock Stream	84	Mareen's Brook	65
Brook Stream (Skerries)	5	Frescati Stream	81	Marino Stream	37
Brook, The	5	Furry Glen Stream	50	Matt River	4
Brownsbarn Stream	53	Gallanstown Stream	54	Maureen's Brook	65
Cabinteely Stream	84	Gallblack Stream	54	Mayne River	18
Camac, River	52	Garristown Stream	2	Mickey Brien's Stream	82
Cammock (Cammack) River	52	Gaybrook Stream	15	Middle Brook	89
Camoke, Cammoke River	52	Glanarawny	53	Mill Stream	5
Cappoge Stream	32	Glanduff	89	Milverton Stream	5
Carrickbrack Stream	23	Glas Naion	35	Monkstown Stream	
Carrickbrennan Stream		Glas Nevin	35	(Carrickbrennan)	82
(Rochestown)	82	Glascoynock	56	Monkstown Stream (Stradbrook)	82
Carrickbrennan Stream		Glasholac	58	Mount Merrion Stream	81
(Stradbrook)	82	Glaslower	81	Mount Olivat Stream	33
Carrickmines River	85	Glasnabrockey	89	Moyne River	18
Carysfort-Maretimo Stream	81	Glassavullaun (Stream)	67	Muckross Stream	75
Castle Stream	74	Glasthule	83	Naniken (Nanekin) River	26
Cataract of the Brown Rowan		Glebe North Stream	4	Narahan Water	53
(Roan) (Tree)	65	Glen River	73	Nine-Stream River	8
Cemetery Drain	35	Glenamuck Stream	85	Nutley Stream	80
Churchtown Stream	74	Glenaulin Stream	50	O'Toole's Stream	83
Clarade	80	Glencullen River	89	Offington Stream	21
Claremont Stream	35	Glendoo Stream	71	Oldcourt Stream	68
Clondalkin River	53	Glenmunder Stream	88	Olney Stream	76
Clonee Stream	30	Glin, River	72	Orlagh Stream	70
Clonkeen Stream	83	Golf Stream	85	Owendoher River	71
Clonmethan Stream	7	Grallagh Stream	8	Owent(h)rasna (River)	71
Collinstown Stream	6	Grange Stream	19	Palmerstown Stream	6
Commons Water	61	Grange Stream (Whitechurch)	72	Pill, River	9
Cookstown River	89	Gray's Brook	21	Pinkeen Stream (Eastern)	31
Coolfan River	53	Griffeen River	49	Pinkeen Stream (Western)	30

96

The Rivers and Streams of Dublin (City of Dublin, Fingal, South Dublin, Dún Laoghaire-Rathdown)

Piperstown Stream	67	Skillings Glas	26	Toberach	53
Poddell	58	Slade (More) River (Camac)	52	Tobermaclugg Stream	49
Pole Water	56	Slade Brook	66	Tolka, River	30
Portrane (*Portraine*) Stream	8	Slang, River	74	Toulchy Water	30
Pottle, River	58	Sluice River	17	Toulghy, River	30
Pound Lane Stream	50	Sohlang River	73	Trimleston Stream	79
Priory Stream	81	Soulagh, River	58	Tromanallison	65
Puddell	58	St Brigid's Stream (Phoenix Park)	50	Trumandoo	65
Quinn's River	25	St. Bride's Stream	84	Tulechan	30
Racecourse Stream	85	St. Catherine's Stream	6	Tullaghanoge, River	30
Racreena River	49	St. Helen's Stream	79	Turnapin Stream	18
Raheny River	26	St. Laurence Stream	51	Turvey River (*Stream*)	9
Rathmooney Stream	6	St. Margaret's Road Stream	33	Two Slades, The	53
Regles Stream	7	St. Margaret's Stream	12	Tymon River	58
Richardstown River	8	St. Michan's Streams	56	Viceregal Stream	55
River Poddle	58	Staffordstown Stream	9	Wad River	28
Robinhood Stream	53	Stayne River	63	Wad Stream	17
Roche, The (River)	2	Stein River	63	Walkinstown Stream	54
Rochestown Stream	82	Steyn River	63	Walshestown Stream	4
Rush (Town) Stream	6	Steyne River	63	Ward River	11
Salagh, River	58	Strad Brook	82	Warrenstown Stream	31
Salmon Stream	4	Stradbrook Stream	82	Westmanstown Stream	50
Sandyford Stream	81	Stream of the Neighing of Horses	53	Whitechurch Stream	72
Santa Sabina Stream	23	Swan River	76	Whitestown Stream	68
Santry River	25	Swan Water	76	Whitewater Brook	22
Saucerstown Stream	11	Swift River (Brook)	52	Wimbletown Stream	8
Scribblestown Stream	32	Swords River	11	Wood Brook	87
Seagrange Park Stream	19	Tallaght Stream	68	Woodtown Stream	71
Shallon Stream	12	Tanner's Water	4	Wyckham Stream	74
Shanganagh River	84	Thomastown Stream	83	Zoo Stream	55
Skerries Stream	5	Ticknock Stream	74		

Constructed watercourses

Abbey Stream	60	*(Old) City Watercourse, The*		*Limerick Watercourse*	63
Artificial Watercourse, The	66	*(Stone Boat to St James's)*	60	*Lord Limerick's Watercourse*	63
Back Course, The	63	*Col(e)man's Brook*	62	*"Middle Arch watercourse"*	37
Camac Millrace	54 & 62	*Crockers' (Barrs) Stream*	62	*Tenter Water*	61
City Conduit, The	62	*Glib Water*	62	*Thomas Street Watercourse*	62
City Watercourse, The (Dodder		*Grace Park Watercourse*	37	*Usher's Island Millrace*	54 & 62
to Poddle / southern)	58	*Lakelands Overflow*	59	*Wad River Diversion*	28 & 35
		Lea Brook	62		

List of named Liffey tributaries & reaches in Counties Wicklow & Kildare

(**County Wicklow**)		Dealbog Brook	39	Ballinagee River	40	Tobenavoher River	43
Uisce an Théir (LIffey reach)	38	Tramhongar	39	Glasnagollum Brook	40	Hartwell River	43
Tromán Ata	38	Luggcullen Brook	39	Gowlan Brook	40	Paintestown River	43
Miley's Brook	38	Ballinatone Brook	39	Glasnadade Brook	41	*Rathmore Stream*	43
Tromán Mór	38	Scurlock's Brook	39	Glenreemore Brook	41	Slane River	44
Cruckan Brook (Liffey reach)	38	Shankill River	39	Asbawn Brook	41	Kill River	44
Meadow Brook	38	Seechon Brook	39	Knockalt Brook	41	Pausdeen Stream	44
Asnabarney	38	The Slade	39	Douglas River	41	Toni River	44
Quarry Brook (tributary /		Cloghoge Brook	39	Trether Brook (I)	41	Crodaun (Croudaun)	
Liffey reach)	38	**Brittas River*** (also in		Trether Brook (II)	41	Stream	45
Grace's Brook	39	Co. Dublin)	39	Little Douglas Stream	41	Donaghcumper Stream	
Lugnalee Brook	39	**Kilbride River** (part of *)	39			(or *River*)	45
Askakeagh	39	**Lisheens River** (part of *)	39	(**County Kildare**)		Shinkeen Stream	45
Carriglaur	39	Ballyward Brook	40	Seasons Stream	41	Barnhall Stream	45
Carrigvore	39	Woodend Brook	39	Lemonstown Stream	41	Ryan's Stream	45
Sraghoe (Sraughoe)	39	Black Brook	39	Toor Brook	41	**Rye Water** (*Ryewater*	
Glossnavillogue	39	Dwyer's Brook	39	Brook of Donode	41	*River, Rye River*)	46
Glasnaslingan	39	*Srothansoillaghe*	40	Kilcullen Stream	41	Lyreen River	46
Cransillagh	39	Tromawn Brook	40	*Mill Stream*	41	*Loughtown River*	46
Athdown (Adown) Brook	39	Ballynastockan Brook	40	Pinkeen Stream (Kilcullen)	41	Baltraccy River	46
Shaking Bog Brook	39	Cock Brook	40	Sexes Stream	41	Clonshanbo River	46
Ballylow Brook	39	Fraughan Brook	40	Mooney's Stream	41	Joan Slade River	
Lugduff Brook	39	Oghill Brook	40	Pinkeen Stream		(Owenslade)	46
Lavarnia Brook	39	Troman Brook	40	(Morristown)	42	*Meadowbrook River*	46
Ballydonnell Brook	39	Ballyknockan Brook	40	Awillyinish (Annislingh)		*Rowanstown River*	46
Boleyhemashboy Brook	39	Annacarney Brook	40	Stream	42	Glashnoonareen River	47
Glenvadda Brook	39	Quainty Brook	40	Butter Stream (*Butterstream*)	42	*Offaly River*	47
Whitebog Brook	39	**Kings River** (Owenree)	40	Gollymochy River	42	Silleachain Stream	48
Glenagoppul Brook	39	Annalecka Brook	40	Morell River	43		
Parkbawn Brook	39			Annagall Stream	43		

(from top): Main outfall of Bride's Stream west of Rush / Broad Meadow Water (Broadmeadow River) / the Liffey comes down from Leixlip Dam to meet the Rye Water / the Rye Water in turn approaching the Liffey at Leixlip, passing Leixlip Castle grounds / the Poddle comes out of culvert between Mount Jerome Cemetery and the Russian Orthodox church complex / the last curve of the Mill Stream (Skerries) towards Skerries Strand / the Little Dargle River by the former site of Phibb's Weir, where a diversion to the Wyckham Stream and on to the Dundrum Slang River used to be taken off.

(Historical, clockwise from above) Two Liffey city quay scenes; Kavanagh painting (probable) of the Santry mouth, Raheny; the former Salmon Leap at Leixlip (to which the town's name refers); the Strawberry Beds in the Liffey Valley, once a popular picnic spot (Diswellstown Stream flows to the Liffey nearby); Dublin from the hills of the Phoenix Park; the (no longer extant) Poulaphuca Falls on the Liffey; Castletown House seen from across the Liffey; historical scene with the Rye Water confluence and Leixlip Castle's fancy boathouse).

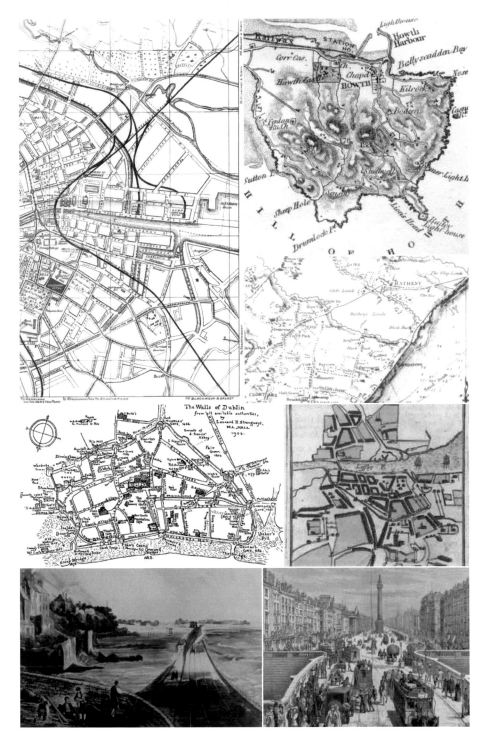